"Neil Bradbury writes with wit, flair, and authority in a debut that will both shock and delight readers. *A Taste for Poison* is a perfect mixture of pop science, true crime, and medical history. I loved this book, and so will you."

—Lindsey Fitzharris, bestselling author of *The Butchering Art: Joseph Lister's Quest to Transform the Grisly World of Victorian Medicine*

"A taut, exciting read about some of the most diabolical killers in history—the poisoners. Bradbury's fusion of science and slayings is grounded in research that reads like a thriller."

—Kate Winkler Dawson, author of *American Sherlock: Murder, Forensics, and the Birth of American CSI*

"*A Taste for Poison* is a delightfully detailed and engaging exploration of eleven poisons. It is a brilliant mix of well-known and less familiar poisons that draws on a huge range of real-life cases and examples. Even those already familiar with poisoning cases or physiology will find something new and intriguing in its pages. Bradbury's enthusiasm for his subject shines through. A fascinating read from start to finish."

—Kathryn Harkup, bestselling author of *A is for Arsenic: The Poisons of Agatha Christie*

PRAISE FOR *A TASTE FOR POISON*

"Bradbury takes the reader on a lively spin through the histories of eleven of the most commonly used poisons and toxins, using real cases—famous and otherwise—to explain how each works in the body. A professor of physiology and biophysics, Bradbury is an engaging, cheerful tour guide."

—*The New York Times Book Review*

"Macabre? Undoubtedly. Mesmerizing? Exceptionally . . . True crime enthusiasts will enjoy this erudite and entertaining book."

—*The Free Lance-Star*

"An accessible and fascinating study of poisons, using real murder cases to explain how the chemicals affect the human body. Readers of *A is for Arsenic: The Poisons of Agatha Christie* will be entertained."

—*Publishers Weekly* (starred review)

"Appealing to any true crime fan, *A Taste for Poison* is a genre-bending book that holds readers' interests through each chapter."

—*Booklist*

"Bradbury takes readers on a frightening romp . . . fascinating, edifying, and terrifying. This absorbing volume about murderers' use of poison will appeal to true crime lovers and fans of popular science in the vein of Mary Roach."

—*Library Journal*

"A fascinating tale of poisons and poisonous deeds that both educates and entertains."

—Kathy Reichs, author of the Temperance Brennan *Bones* series

"*A Taste for Poison* weaves jaw-dropping true stories and spellbinding histories behind the most infamous poisons in the world. Even some of the slightly lesser-known substances, such as potassium and digoxin, come alive on the page as both notoriously lethal and yet rather helpful, depending on their dose. Page after page, readers will be alternately entertained, horrified, and educated in turn on these notorious 'molecules of death.' A thrilling nonfiction read!" —Lydia Kang, bestselling author of *Quackery: A Brief History of the Worst Ways to Cure Everything*

"Pick your poison—valued as a curative or engineered for war. Neil A. Bradbury uses true crime and real science to explore both the history and the biochemistry of how toxins work and how they kill. *A Taste for Poison* is an unselfconsciously jaunty work of horror, and its stories may leave you eyeing your housemates and sniffing your coffee." —Judy Melinek, M.D., and T. J. Mitchell, *New York Times* bestselling coauthors of *Working Stiff: Two Years, 262 Bodies, and the Making of a Medical Examiner*

"Over the centuries, as poisoners became more able at their craft, scientists became more able to explain how and why poisons work. . . . Despite the descriptions of horrifying deaths in horrifying detail, I couldn't stop reading *A Taste for Poison*. I just had to find out how and why forensic sleuths manage to solve what once might have been considered a perfect crime."
—Penny Le Couteur, bestselling coauthor of *Napoleon's Buttons: How 17 Molecules Changed History*

A Taste for Poison

ELEVEN DEADLY MOLECULES AND THE KILLERS WHO USED THEM

Neil Bradbury, Ph.D.

ST. MARTIN'S GRIFFIN
NEW YORK

Published in the United States by
St. Martin's Griffin, an imprint of St. Martin's Publishing Group

www.stmartins.com

Designed by Michelle McMillian

The Library of Congress has cataloged the hardcover edition as follows:

Names: Bradbury, Neil, author.
Title: A taste for poison : eleven deadly molecules and the killers who used them / Neil Bradbury.
Description: First edition. | New York : St. Martin's Press, [2021] | Includes bibliographical references.
Identifiers: LCCN 2021026597 | ISBN 9781250270757 (hardcover) | ISBN 9781250270764 (ebook)
Subjects: LCSH: Poisoners—Case studies. | Poisoning—Case studies. | Poisons—Physiological effect—Case studies.
Classification: LCC HV6549 .B73 2021 | DDC 615.9—dc23
LC record available at https://lccn.loc.gov/2021026597

ISBN 978-1-250-62451-2 (trade paperback)

First St. Martin's Griffin Edition: 2023

10 9 8 7 6 5 4 3 2 1

To my wife and daughters, and to my parents
for teaching me right from wrong

Contents

As a rule, women are the great poisoners, although I do recall with pleasure the case of the gentleman solicitor in Wales who poisoned everybody in sight. He couldn't stop himself. He was very genteel. He came up with the most memorable line in the annals of true murder. As he handed one of his guests a poisoned scone, he said, "Excuse fingers."[1]

—SIR JOHN MORTIMER, BARRISTER, AUTHOR AND CREATOR OF *RUMPOLE OF THE BAILEY*

PART I

Biomolecules of Death

Introduction

I love the old way best, the simple way of poison,

where we too are strong as men.

—Euripides, *Medea,* 431 BC

Within the annals of crime, murder holds a particularly heinous position. And among the means of killing, few methods generate such a peculiar morbid fascination as poison. Compared with hot-blooded spur-of-the-moment murders, the planning and cold calculations involved in murder by poison perfectly fit the legal term *malice aforethought*. Poisoning requires planning and a knowledge of the victim's habits. It requires consideration of how the poison will be administered. Some poisons can kill within minutes; others can be given slowly over time, gradually accumulating in the body but still leading inexorably to the victim's death.

This book is not a catalog of poisoners and their victims, but rather explores the nature of poisons and how they affect the body at the molecular, cellular, and physiological levels. Each poison kills in its own unique way, and the varied symptoms experienced by the victims often give clues as to the nature of the poison used against them. In a few instances such knowledge has led to appropriate treatment and full recovery. In other cases knowledge of the poison is not of therapeutic benefit, because there is simply no antidote.

The words *poison* and *toxin* are often used interchangeably, though strictly speaking they are not the same thing. Poisons are any chemicals that cause harm to the body, and can be natural or man-made, whereas toxins usually refer to deadly chemicals made by living things. If you are on the receiving end of either though, the difference is somewhat academic. The word *toxikon* comes from ancient Greek, meaning "a poison into which arrows are dipped," and describes plant extracts smeared onto arrowheads to induce death. When the word *toxikon* was combined with the Greek term *logia*, meaning "to study," we ended up with toxicology, or the study of toxins. The word *poison* is derived from the Latin word *potio*, which simply means "drink." This slowly morphed into the Old French word *puison* or *poison*. The first appearance of *poison* in the English language appeared in 1200, meaning "a deadly potion or substance."

Poisons obtained from living organisms are often mixtures of many chemicals. For example, crude extracts from the deadly nightshade plant (also known as belladonna) are quite dangerous, but from these extracts comes the purified chemical atropine. Similarly, foxglove plants are also poisonous, but the single chemical digoxin can be purified from the plant.

Historically some poisons have been created by mixing

together several different poisons, as seen in *aqua tofana,* a mixture of lead, arsenic, and belladonna.[1]

How does a chemical sitting in a bottle doing no harm to anyone end up as a poison inside a dead body? Whatever the poison may be, there are three distinct stages that occur before death: delivery, actions, and effects.

Poison can be delivered to a victim through four routes: ingestion, respiration, absorption, or injection. That is, they may be eaten or drunk and enter the body through the intestine; inhaled into the lungs; absorbed directly through the skin; or injected into the body, either into muscles or the bloodstream. How a killer gets the poison into a victim's body depends on the nature of the poison. Although poisonous gases have been used to kill, they involve a degree of technical difficulty that makes them impractical to utilize, and often makes it hard to target a specific individual. Absorption through the skin or mucous membranes of the eyes and mouth can be quite effective: The killer does not have to have any contact with the victim or even be around when the poisoning takes place. Simply smearing the poison on something the victim will touch can be sufficient to cause death. Mixing with food or drink provides an easy route for most poisons. This works especially well for solid crystalline poisons that can simply be sprinkled onto a meal or dissolved in a drink. However, some poisons must be injected into the body. Sometimes this is because the poison is a protein that would simply be broken down by the stomach and intestines if it was eaten. Of course the killer must be close enough to the victim to inject the poison.

Now we turn to the crux of poisons: How do they disrupt the inner workings of the body? Exactly what poisons do is incredibly varied, and their actions reveal much about human biology. Many poisons attack the nervous system, disrupting the highly sophisticated electrical signaling that controls the normal functions of the body. Undermining the communication between parts of the heart can easily be seen to stop the heart from beating and cause death. Interrupting the regulation of the diaphragm, the muscle that controls breathing, can similarly cause death by shutting down respiration and causing asphyxia. Other poisons get inside the cells of the body by pretending to be something that they are not. Disguising their true nature, and having almost, but not quite, the same shape as a vital component of the cell, these poisons can be incorporated into the cell's metabolism but are unable to perform the right biochemical functions. With the poison acting as a counterfeit molecule, the whole of the cell's chemistry grinds to a halt and the cell dies. When enough cells die, so does the whole body.

It is not too hard to imagine that if different poisons work in different ways, the symptoms experienced by the victims would also be different. For most ingested poisons, irrespective of how they work, the first response is often vomiting and diarrhea, in a physical attempt to remove the poison from the body. Poisons that affect the nerves and electrical signaling of the heart will be experienced as heart palpitations and, sooner or later, cardiac arrest. Poisons that affect the cells' chemistry often cause nausea, headaches, and lethargy. It is stories of the actions of poisons and their dreadful consequences that fill the rest of this book.

While most people would consider poisons to be lethal drugs, scientists have used the exact same chemicals to tease apart the inner molecular and cellular mechanisms of cells and organs, using this information to develop new drugs that treat and cure a wide range of diseases. For example, studying how the poisons in the foxglove plant affect the body has led to the development of drugs to treat congestive heart failure. Similarly, understanding how belladonna affects the body has helped create drugs now routinely used in surgery to prevent postoperative complications, and even to treat soldiers exposed to chemical warfare. From this it can be seen that a chemical is not intrinsically good or bad, it's just a chemical. What differs is the intent with which the chemical is used: either to preserve life—or to take it.

. { 1 } .

Insulin and Mrs. Barlow's Bathtub

Both Williams of Rochester and Woodyatt of Chicago had patients who died of hypoglycemic shock after receiving an overdose of insulin.

—Thea Cooper and Arthur Ainsburg, *Breakthrough*, 2010

MIRACLE DRUG TO MURDER WEAPON IN THIRTY YEARS

What does the word *poison* conjure in your mind? Extracts from poisonous plants, toxins from venomous snakes, or maybe mad scientists making deadly chemicals deep in an underground bunker? Not all poisons have such exotic pedigrees. Sometimes what makes things toxic is exactly what allows them to be used for good.

This apparent contradiction between a chemical being both toxic and tonic was first appreciated in the medical revolution during the Renaissance. Philippus Aureolus Theophrastus Bombastus von Hohenheim (fortunately better known by his nickname Paracelsus), the great sixteenth-century alchemist and physician, cautioned: "*It is the dose that makes the poison.*" Perhaps nowhere is there a better example of this than in our first poison: a chemical that has lifesaving properties in small doses but is deadly when applied in large amounts.

The chemical in question is insulin, and its absence, or an inability of the body to respond to it properly, leads to the disease diabetes mellitus.[1] Before the widespread availability of insulin, a diagnosis of diabetes was akin to a death sentence. The most optimistic prognosis was a few years of suffering, followed by death. Diabetes would transform a happy active childhood into one of ravenous hunger and insatiable thirst. In the decade before insulin's discovery, the American doctors Frederick Allen and Elliott Joslin advocated severe fasting to prolong the lives of diabetic patients. It was a depressing process of slow starvation, with patients whittled down to little more than skin and bones.[2] It was known that diabetics had sugar in their urine, and starvation certainly stopped that from happening. However, this approach was really just treating the symptoms, and there was little scientific evidence to support the diet as a viable therapy—but neither was there a rational alternative.

Things changed in 1921, when Canadian researchers managed to identify and purify insulin from animal pancreases. The first patient treated with insulin was fourteen-year-old Leonard Thompson, a boy weighing only sixty-five pounds and drifting in and out of a diabetic coma. With insulin treatments, Leonard's blood sugar levels fell dramatically toward normal, he started to regain weight, and his symptoms gradually disappeared. Although not a cure, insulin injections allow the millions of individuals with diabetes to live a full, and reasonably normal, healthy life. One of the most important things all patients with diabetes are taught is to recognize the symptoms of too little and too much insulin.

It was a remarkably short time from the initial discovery and purification of insulin to its widespread use in treating patients with diabetes, with commercial insulin available in 1923, only two years after insulin's discovery.[3] A more sinister and tragic timetable shows that it took barely three decades to turn a lifesaving chemical into a deadly murder weapon.

MRS. BARLOW'S BATH

Detective Sergeant John Naylor was called to a semidetached house on Thornbury Crescent in Bradford, England, in the early hours of Saturday, May 4, 1957. As he entered the house, Naylor heard faint sobbing and found a distraught husband in the throes of grief, tightly gripping a picture of a woman. Naylor was directed to the upstairs bathroom by a police constable, and there, slumped in the bath, naked and dead, was the woman in the picture. Anxious neighbors stood close to the weeping husband in uncomfortable silence, convinced that his grief was genuine—but Naylor was not so sure.

To all who knew her, Elizabeth "Betty" Barlow seemed happily married to her devoted husband, Kenneth. According to neighbors, they were extremely happy together and never argued.

Elizabeth, nine years younger than Kenneth, was actually Barlow's second wife, having married him in 1956 following the death of his first wife. In marrying Kenneth, Elizabeth also became stepmother to Barlow's young son, Ian. Both Kenneth and Elizabeth had worked in various hospitals around the Yorkshire town of Bradford, Elizabeth as an auxiliary nurse, and Kenneth as a state registered nurse, which may have been how the couple first met.

After the wedding Kenneth continued working as a nurse at Bradford Royal Infirmary, but Elizabeth left nursing and took a job in the ironing section of a local laundry. The work was fairly mundane, and clouds of steam constantly swirled around her, making her clothes damp and uncomfortable; but the pay was reasonable, and the job helped with the family finances. Fridays were Elizabeth's half-day at work, and Friday, May 3, 1957, was no different. Noon was approaching, and as Elizabeth eagerly gathered her things to leave work, she mentioned to her friends that she was looking forward to a little time to herself so that she could wash her hair. On the short walk from the laundry to their home in Thornbury Crescent, Elizabeth stopped at the local fish-and-chip shop to pick up lunch for the family. At twelve thirty the hot fish and chips were unwrapped from their vinegar-soaked newspaper coverings and plated with bread and butter, all washed down with cups of tea.

After lunch Elizabeth busied herself doing some housework and washing the family's clothes, while Kenneth spent his Friday afternoon taking care of his pride and joy by pulling the car out of the attached garage and giving it a thorough wash. Around four o'clock, Elizabeth went to visit Mrs. Skinner, her next-door neighbor, who would later testify that Elizabeth appeared cheerful and "full of life." "As a matter of fact, she showed me a set of black underwear she [bought], and joked about it," Mrs. Skinner recalled.

That evening the family moved to the living room to relax. Elizabeth lay on the sofa for a short time but became increasingly restless, eventually telling her family she was going to go lie down for a while. At 6:30 p.m., as she climbed the stairs, Elizabeth called out to Kenneth asking him to come get her in an hour, as she wanted to watch a television program with him. As it turned out, Elizabeth would never watch television again. Fifty minutes later Kenneth climbed the stairs to let his wife know that the show was about to start, but Elizabeth had already changed into her pajamas and gotten into bed, telling her husband that she felt "too comfortable to move." Kenneth ambled back to the living room alone to watch television for the next half hour, before taking a glass of water up to his wife to see how she was doing.

In the bedroom Kenneth found Elizabeth still in bed and feeling very tired. He would later testify that his wife told him that she was "too tired to say goodnight to her stepson." It was still a bit early for Kenneth to retire for the evening, and he wanted to give his wife some time alone to rest, so he went back downstairs to finish watching television. Shortly before nine thirty, Kenneth heard Elizabeth calling him from their bedroom. As he climbed the stairs and entered their bedroom, Kenneth was disturbed to find that his wife had vomited on the bed. The couple changed the sheets, and Kenneth took the soiled bed linens downstairs, where he placed them in the kitchen washtub. Not only was

Elizabeth complaining of being tired, but she was now "feeling too warm," and decided to lie on top of the newly made bed-covers.

Kenneth changed into his pajamas, got into bed, and started reading. By ten o'clock Elizabeth was still not feeling well and was now sweating profusely. She undressed and told her husband she was going to take a bath to try to cool down. Kenneth heard the bathwater running before he dozed off to sleep.

Suddenly something startled him awake. Glancing at the alarm clock on the bedside table, he saw it was already 11:20 p.m. and was surprised that his wife had not yet finished her bath and returned to bed. He anxiously called out to Elizabeth, asking "Is everything all right? How much longer are you going to be in there?" He got no reply. Worried that she had fallen asleep in the now-cold bathwater, Kenneth got out of bed and went to the bathroom, where to his horror, he found Elizabeth totally submerged in the water and not moving.

In a panic, Kenneth was convinced that his wife was drowning, and quickly pulled the plug to let the water out of the bath. As it drained, Kenneth desperately tried to pull his wife from the bathtub onto the hard bathroom floor, but try as he might, he just couldn't lift her out. Fortunately, as a trained nurse, Kenneth realized that he would have to perform artificial respiration while his wife was still in the bathtub. In vain he tried coaxing air into Elizabeth's lifeless lungs, but he needed help.

Without a telephone in their own house, Kenneth rushed next door, still dressed only in his pajamas, rousing his neighbors, the Skinners. Anxiously Barlow begged them to fetch a doctor while he went back to again try to resuscitate his wife. Strangely, the neighbors, rather than immediately calling for an ambulance, decided to see for themselves what was going on. As they walked next door and up the small flight of stairs to the bathroom, they were shocked to find Elizabeth's naked body still lying in the empty bathtub, Kenneth rubbing her shoulders. With the Skinners now convinced of the seriousness of the situation, a telephone call was made to the family doctor, imploring him to get there as soon as he could. As they waited for the doctor, Mrs. Skinner glanced over at Kenneth, who was sitting in an armchair, his face buried in his hands, gently sobbing. Despite the prompt arrival of the doctor, it was too late for Elizabeth, and she was pronounced dead.

Death is always unsettling, but even more so when the deceased is a young wife and mother, otherwise in perfectly good health. The doctor couldn't quite put his finger on it, but something just didn't seem quite right. Certainly Elizabeth was dead, and the telltale signs of rigor mortis were beginning to set in, but a gnawing in his gut convinced him that he should talk to the police, and it was not long before Detective Naylor arrived to examine the scene.

Elizabeth's decision to take a bath that night would indeed turn out to be pivotal. Had she stayed in bed, it is likely that her death, though distressing in one so young, would have been ruled as natural. It seemed at first as though Elizabeth had drowned, but her pupils were markedly dilated, far more than the doctor had ever witnessed in a drowning victim.

But what had caused Elizabeth's pupils to dilate? What had

made her so hot that she needed a cold bath to cool down? And what had made a young vibrant woman so tired? Remarkably, the answer to Elizabeth's demise centered around something very simple, and something millions of people put into their coffee and tea every day: sugar.

"JUST A SPOONFUL OF SUGAR . . ."

What we buy in grocery stores is only one kind of sugar. Sugars are known chemically as carbohydrates, because they are all made from carbon, hydrogen, and oxygen atoms, linked together in particular ways. The smallest sugars are made from just six atoms each of carbon and oxygen and twelve of hydrogen. Depending on the arrangement of the atoms, they either make fructose (or fruit sugar), galactose (found in foods like milk and avocados), or glucose. When we talk about "blood sugar" we really mean glucose, which is transported in the blood as an energy source. The white crystals we spoon into coffee and tea as "sugar" are really table sugar, or more properly sucrose, which is made by joining together a molecule each of fructose and glucose. Similarly milk sugar, or lactose, is glucose and galactose molecules joined together.

It's also possible to link hundreds and thousands of carbon, oxygen, and hydrogen molecules together in long sugar chains to form glycogen in animals or starch and fiber in plants.[4]

One of the remarkable things about the body is that no matter what carbohydrates we eat, be it American French fries (British chips), bread, or pasta, or sugar in sodas and fruit juice, all the carbohydrates are broken down in our intestines to the three sugar building blocks of glucose, fructose, and galactose, to be absorbed and shipped off to the liver. There the different kinds of sugars are converted into glucose, making glucose the only sugar transported in the blood.

Like many substances in the body, glucose levels in the blood are kept within relatively narrow limits. If they wander too far from these preset boundaries, severe complications and even death can result. With too little glucose in the blood (hypoglycemia), there is not enough to supply the energy needs of the body (particularly the brain), but with too much glucose (hyperglycemia), damage to delicate cell membranes, particularly those of the nerves and the retina, can occur, leading to nerve damage and pain or even loss of eyesight. Unlike other organs in the body, the human brain uses glucose as its primary fuel supply. Since the brain has no way to store glucose, nerves in the brain are critically dependent upon a constant and steady supply from the blood to function properly. If blood glucose falls below 50 percent of normal levels, the fingers and lips go tingly and numb, the brain becomes sluggish, and thinking can be confused and unfocused. Sweat begins to bead over the body; the heart pounds, trying to circulate glucose that is no longer in the blood. The voice slurs and vision blurs. At 25 percent of normal levels, coma and even death can ensue.

Given the serious consequences if blood glucose levels fall too much or too rapidly, it is not surprising that the body has a way to carefully control and regulate the amount of sugar in the blood: a hormone called insulin.

INSULIN AND BLOOD SUGAR LEVELS

What was it about Mrs. Barlow's death that suggested foul play? To understand her symptoms, we need to look at the role of insulin in regulating the level of glucose in the blood. Located close to the liver and just below the stomach is the pancreas, an organ roughly the shape and size of a banana. The pancreas performs a number of essential functions in the body, including releasing enzymes into the gut to aid the digestive process. But the pancreas also produces insulin, the hormone that helps the body store and use glucose. When a meal containing carbohydrates has been digested, the levels of glucose in the blood rise, which triggers the pancreas to release insulin into the bloodstream. Once released from the pancreas, insulin is transported to the key organs of the liver, adipose or fatty tissue, and muscle.

When these organs are exposed to insulin, they ramp up their ability to absorb glucose from the blood. Because of this quick absorption, even after consuming a meal with lots of sugary foods, the level of glucose in the blood rises for only a short time before coming back down to a normal resting level. In this way insulin performs two critical functions in the body: first, to prevent blood glucose levels from getting too high; and second, to cause the liver, muscles, and adipose tissue to take excess glucose from the blood. The liver and muscle store it as glycogen; the adipose tissue, as fat. As blood glucose levels drop, so too does the level of insulin released by the pancreas. But what would happen if insulin levels failed to go back down, and the pancreas kept releasing insulin into the blood? What would happen if the signal to the liver, muscle, and adipose tissue to remove glucose from the blood was never turned off? Fortunately this never happens naturally, except in some very rare cancers. But what if insulin levels were artificially elevated—say, by someone injecting large doses of insulin into the bloodstream? This was the question a young doctor working in Berlin in the early twentieth century was asking, in the hopes that he might be able to help some of his patients.

INSULIN SHOCK AND CLUES TO ELIZABETH BARLOW'S SYMPTOMS

Less than a decade after the first commercial production of insulin, it had become invaluable in treating patients with diabetes. In 1928 an Austro-Hungarian doctor, Manfred Joshua Sakel, was treating a diabetic patient who also happened to have schizophrenia. While trying to control the patient's diabetes, Sakel accidentally gave his patient an overdose of the newly discovered insulin—and was surprised to find that the patient's schizophrenia appeared to have gone into remission. Sakel wondered whether other patients with schizophrenia, who didn't have diabetes, would respond in the same way.

When the patients were injected with insulin, it caused a dramatic drop in their blood glucose levels, depriving the brain of an essential component for normal physiological function. The patients began to sweat profusely, requiring repeated baths to wash away the sweat. As glucose levels fell further, patients became increasingly restless, followed by major convulsions that ended only when they entered into a coma, at which point the patients all had fixed, widely dilated pupils. All these symptoms were on display during Elizabeth Barlow's last hours, as one of the telltale signs that a patient was in a deep insulin coma is fixed, widely dilated pupils. Although the symptoms of schizophrenia, including delusions, hallucinations, agitation, and inappropriate reactions, seemed to abate after insulin shock,[5] no one knew whether the effect was due to the insulin itself or to the coma induced by the insulin.[6] Even though insulin shock seemed to work, the problem remained that for the treatment to be beneficial, the patients had to recover from their insulin-induced comas.

Receiving no support for his research from the hospital where he worked, Dr. Sakel embarked on a series of animal experiments in his own kitchen. Here he convinced himself that a coma induced by low blood glucose (hypoglycemic coma) could be easily reversed through the administration of intravenous glucose. His studies convinced him that he was "on the road to great discoveries."

Sakel left Berlin and returned to Austria to work as a volunteer in Vienna's University Clinic, where he could practice his technique of deep insulin coma therapy (also known as insulin shock therapy) on the clinic's psychiatric population. Since placing patients in an insulin coma was a life-threatening procedure, it was necessary to counteract the effects of insulin by feeding glucose to patients through a rubber tube pushed into the mouth and down into the stomach. If the glucose was not administered promptly, insulin shock therapy was not without its own problems. Depriving the brain of nutrients for extended periods of time produces damage to the cerebral cortex, causing its crenellated architecture to become flat and smooth, much like the brains of people suffering from neurodegenerative disorders. Fortunately, in most cases, Sakel's patients rapidly revived, and there were usually marked signs of psychological improvement.

By 1935 Sakel had published more than thirteen papers on his technique and was reporting an astounding 88 percent success rate in treating psychiatric disorders. Word of his success spread quickly, and Sakel became the darling of the psychiatric community, convincing himself that a Nobel Prize was just around the corner. More and more doctors adopted Sakel's methods, and the technique was slowly implemented throughout Europe and the United States. Doctors giddily competed to see how many times a week they could put their patients into an insulin-induced coma, while others pushed the envelope to see how long their patients could be kept in a coma before reviving them. Experienced therapists bragged of their ability to keep patients in an insulin coma for up to fifteen minutes before administering intravenous injections of glucose or squirting glucose solutions into the patient's stomach.

As the therapy became more widespread, doctors started noting differences among patients' responses to insulin, and even day-to-day variations in the same patient. Despite these findings, all the doctors working on insulin coma therapy reported "tremendous enthusiasm" for the technique. With the onset of World War II, many of the European doctors practicing insulin treatment fled Nazism, further spreading the technique to Allied countries.

But regardless of the enthusiasm of doctors for using insulin to treat psychiatric disorders, it was a house of cards waiting to collapse.

In 1953 Dr. Harold Bourne, an experienced British psychiatrist, published a paper titled "The Insulin Myth," arguing that there was no sound scientific basis for the validity of insulin coma therapy. He maintained that the initial psychiatric diagnoses were most likely flawed and based on unreliable and questionable tests. Bourne also claimed that the results of insulin coma therapy were biased by selecting particular patients and ignoring others. More alarmingly, it became apparent that no two hospitals were performing the coma therapy in the same way, with some clinics inducing a coma for one hour and others for an astounding four hours.

The immediate response to Bourne's concerns was not a "thank you for showing us where we've gone wrong," but rather a wave of condemnation from the medical community. Leading psychiatrists sent letters to medical journals criticizing Bourne, with one stating: "it is clinical experience that counts here, despite all [evidence] to the contrary." It would be another five years before a carefully controlled study on insulin coma therapy would be published, a study that clearly showed, beyond any doubt, that insulin coma therapy was a sham.[7] Given the robustness of the controlled studies, it was hard to criticize their findings, and insulin coma therapy was abandoned, any mention carefully brushed under the carpet.

It is a curious coincidence that the report putting the nail in the coffin of insulin's use to treat psychiatric disorders was published in 1957—just a few short weeks before Kenneth Barlow would use insulin to put his wife in a coffin.

A SEARCH OF THE HOUSE

At 2:00 a.m. on May 4, 1957, Dr. David Price, the Home Office[8] forensic pathologist, arrived at the Barlow house to start an examination of Elizabeth's body. Dr. Price's suspicions were aroused initially because the drowning of an otherwise healthy middle-aged female in a home bathtub is extremely rare. Perhaps more telling was the presence of a small cupful of water that remained in the hollow created between Elizabeth's arm and the side of the bathtub. How had that small amount of water lain undisturbed if Kenneth had tried to drag his wife out of the tub, as he'd claimed? If this part of Kenneth's story was implausible, it brought suspicion on his whole version of the night's events. The police scoured every room in the house, finding the vomit-stained sheets and Elizabeth's sweat-soaked pajamas in the sink. In the kitchen, on a shelf above the door leading to a pantry, was a small porcelain pot. Inside the pot, wrapped in a handkerchief, were two used syringes and four hypodermic needles; however, no empty vials of medication were found.

At 5:45 a.m., Mrs. Barlow's body was taken from the house and transported to a local mortuary, where Dr. Price began his postmortem examination. Bloodstained froth in the nose, mouth, and throat, as well as fluid in the lungs, confirmed the initial suspicion that Elizabeth had indeed died from drowning. But why were there no signs of a struggle? No other abnormalities were found, but it was discovered that Elizabeth was eight weeks' pregnant. Blood and urine samples were collected and sent to the North-Eastern Forensic Science Laboratory, but none of the typical poisons or abortion-inducing chemicals were found in any of Elizabeth's samples. Dr. Price became convinced that Elizabeth had been rendered unconscious before being drowned. Based on his knowledge of the recently discredited insulin coma therapy and her telltale dilated pupils, Dr. Price was sure that Elizabeth had been injected with insulin to make her comatose before she was pushed underwater. A crucial question remained: If Elizabeth was injected with insulin, where were the injection marks?

On May 8, four days after Elizabeth's death, and just hours before her funeral, the decision was taken to reexamine her body. Now knowing that they were looking for needle marks, Dr. Price and his team carefully went over the body inch by inch with magnifying glasses. And there they were: two hypodermic injection sites in each buttock. Samples of the injection sites and surrounding tissue were taken, but other than cataloging them and placing them in storage, nothing was done.

During police questioning Kenneth Barlow was confronted with the evidence of syringes in the kitchen and injection sites in Elizabeth's buttocks. Kenneth confessed that he had indeed injected Elizabeth, but it was with her consent, and that it was not insulin, but rather ergometrine, a drug used in obstetrics to prevent heavy bleeding after childbirth by causing contractions of the uterus. It could also be used to induce abortions, which in itself was a criminal offense.

Kenneth told the police that neither he nor Elizabeth wanted another child, and that she had told him she would sooner put her head in a gas oven than bear a child. With no alternative, Kenneth had colluded in the ergometrine-induced abortion attempts. However, Kenneth was unaware that this possibility had been considered by the forensic team and ruled out because of the lack of any ergometrine in Elizabeth's system, and the lack of the drug in either of the syringes. Moreover, ergometrine could not have caused the dilated pupils, sweating, or vomiting that Elizabeth had experienced.

The police were now convinced that Kenneth Barlow had murdered his wife by injecting her with a massive dose of insulin, sending her into an insulin-induced coma from which she would never recover. There was just one more thing that the prosecution needed to bring the case to trial: forensic evidence of high levels of insulin in Elizabeth's body. The problem was that no one had ever measured insulin in human tissue before. Would Kenneth get away with murder for lack of a critical piece of evidence?

CONVULSING MICE AND SLEEPY GUINEA PIGS CONVICT KENNETH BARLOW

Kenneth may have had more than a few reasons to murder his pregnant wife. Given their finances, another mouth to feed would stretch the family's meager income too far. Perhaps Kenneth felt that a new baby was just too inconvenient at his stage of life, and convinced his wife that an abortion before anyone noticed the pregnancy would surely solve all their problems. Since Fridays were Elizabeth's half-day at work, Friday afternoon seemed the best time for the ergometrine injections, giving Elizabeth the weekend to recover.

Sometime after lunch, however, instead of ergometrine, Kenneth injected Elizabeth with a large dose of insulin. Immediately Elizabeth's body responded to the insulin by removing glucose from the blood as her liver, muscles, and adipose tissue absorbed massive amounts of glucose, sending her blood levels plummeting. The fuel necessary to run her brain was draining. As with insulin coma therapy, once a large amount of insulin has been injected and blood glucose depleted, the only remedy is the ingestion of large quantities of glucose. Kenneth was not about to provide that.

A sense of restlessness overcame Elizabeth as she lay on the sofa, a symptom reminiscent of patients undergoing insulin coma therapy. Since there was no energy for her muscles to work, she felt tired, weak, and listless, and so decided to go and lie down in bed. Elizabeth's vomit-soaked pajamas were found in the sink, consistent with the vomiting that often occurs when hypoglycemia sets in. Despite the rather mild weather, Elizabeth was sweating profusely, causing her initially to lie on top of the bed rather than under the covers, and finally leading her to remove her pajamas altogether and seek solace in a bath. With high insulin levels, the pupils of the eye enlarge but remain responsive to light, and Elizabeth's pupils were noted by the pathologist to be so enlarged that he could barely discern the color of her eyes.

As Elizabeth lay in the bath, her brain cells slowly ran out of fuel, sending her into a coma. Once in a coma, did Elizabeth simply slip under the water and drown, or was she held underwater in her unconscious state by a murderous husband? The answer to that question will never be known.

The pathologists had ruled out the presence of ergometrine in Elizabeth's blood, yet they still had not proven the presence of large quantities of insulin in her system. The vomiting, sweating, and dilation of the pupils were all symptoms consistent with hypoglycemia, but this was not objective evidence that would stand up in a court of law. According to today's television crime shows, all the police needed to do was send off some tissue from Elizabeth's body to the local forensics department, and the test results that would persuade a jury to convict would arrive before the credits rolled. Unfortunately, in the late 1950s, forensic science was still in its infancy, and a reliable test for insulin would not be developed for another three years. How, then, could the police prove that Elizabeth had lethal levels of insulin in her body? The police turned to the manufacturers of insulin for help.

Although no one had yet tried to measure insulin levels in human organs—mainly because no one ever thought there would be a need to do so—manufacturers of insulin did need a way to measure the amount of purified insulin they were putting in bottles for diabetic patients, so that patients received the correct dose. The test worked reasonably well when there was a lot of insulin, and the insulin was pure. But would it work when the anticipated levels of insulin were quite small, and far from being pure, the insulin was contaminated by being found in the dead tissues of a potential murder victim?

At the time pharmaceutical companies measured pure insulin using a test delightfully described in the *British Pharmacopoeia* as the "mouse convulsion method of insulin measurement." Mice were injected with doses of pure insulin until their blood sugar was so low that they were unable to sustain normal brain activity, causing them to convulse and go into a coma. Similarly, insulin could also be measured by injecting it into guinea pigs and determining how much was needed to deplete the blood of enough glucose to cause the guinea pigs to "go to sleep."

Although we are now familiar with police laboratories running batteries of tests on samples, police scientists at the time could not be involved with any of the tests because they were not licensed to perform experiments on animals. Fortunately a private firm with a license for animal work agreed to undertake the search for insulin in Elizabeth's tissues. Day after day for weeks on end, they tried to determine how to extract insulin from the tissue removed from Elizabeth's buttocks.

At last a sample was ready, and a tiny amount was slowly injected into a mouse. Immediately the mouse went into convulsions, and a further injection with a glucose solution fully restored the mouse to its initial state. But evidence from one mouse would not sway a jury. Altogether twelve hundred mice, ninety rats, and several guinea pigs were used to determine that a lethal amount of insulin had been present in Elizabeth's body. Convinced that they now had enough evidence for a case against

Kenneth Barlow, a final report was sent to the coroner on July 16, 1957. The cause of death on the charge sheet was: "Asphyxia due to drowning while within a hypoglycemic coma following and due to an overdose of insulin." Kenneth Barlow was arrested for murder and sent to trial in Leeds Assizes court in December 1957.

At trial the prosecution brought forward two witnesses who had had conversations with Barlow some three years earlier, and their recollections were damning. Harry Stork had worked with Kenneth Barlow in a sanatorium where diabetic patients received insulin injections. Stork testified that Barlow told him, "You could commit a perfect murder with insulin. It could not be traced because it dissolves in the bloodstream." The second witness, Joan Waterhouse, was a student nurse at the East Riding General Hospital when Barlow had worked there. She stated that Barlow had informed her, "You could kill somebody with insulin, as it cannot be found very easily in the body unless it is a very large dose." Dr. Price, the prosecution's Home Office expert witness, testified that "Mrs. Barlow died of asphyxia due to drowning while she was in a coma after an injection of an overdose of insulin."

The prosecution argued that Barlow had the motive to commit murder, not wanting to bring up another child to stretch the family's finances. Opportunity for the murder was evident; but what of means?

Miss Ellen Simpson was a senior nurse at St. Luke's Hospital when Barlow had worked there as a nurse. Miss Simpson testified that during his time at St. Luke's Hospital, part of Barlow's duties included giving insulin injections to patients. Barlow had had access to supplies of insulin, and no check was kept on the number of units he "consumed."

Throughout the trial Kenneth Barlow protested his innocence, yet he could provide no reason for the presence of high levels of insulin in his wife's body other than suggesting she had injected herself in the buttocks with insulin.

Naturally the defense lawyers had their own expert witness. Dr. J. R. Hobson argued that the amount of insulin found in Mrs. Barlow's body was perfectly natural. He explained that in moments of stress, like anger or fear, the body automatically pumped adrenaline into the bloodstream, which naturally increased the flow of insulin. Dr. Hobson went on to explain to the jury, "If Mrs. Barlow knew she was slipping down and drowning in the bath and that she could not get out, she would be terrified. . . . That, I think would produce all the symptoms the chemists have described."

In fact adrenaline has the opposite effect on insulin, causing its levels to decrease.

After a grueling five-day trial, the presiding judge instructed the jury that there was only one crime of which Kenneth Barlow could be convicted: "This is murder or nothing," he told them. "If you are satisfied that he injected insulin into his wife, and knowingly injected it, you will probably find no difficulty in reaching the decision that he did so with intent to kill." The jury took eighty-five minutes to find Barlow guilty. Sentencing him to life imprisonment, the judge described Barlow as "a cold, cruel premeditated murderer who, but for a high degree of detective ability, would not have been found out." (The jury was also excused from further service for ten years, because they had listened to "a long and harrowing case.")

As Barlow began his life term, police released new information about the death of his first wife, Nancy. Also a nurse, she had been married to Barlow for twelve years. On May 9, 1956, Nancy suddenly became ill and died a mere twelve hours later. An anonymous telephone call had prompted the police to stop her funeral and order a postmortem, but despite extensive examination, all that was found was some mild brain swelling. The funeral proceeded, and two months later Barlow was married to his second wife, Elizabeth—who would be dead within a year.

Although technically Elizabeth Barlow died from drowning, her insulin-induced coma likely rendered her unable to resist her husband pushing her beneath the water. Kenneth is credited with being the first person to commit murder using insulin as a weapon. In November 1983, at the age of sixty-six and after serving twenty-seven years of his life sentence, Barlow was finally released from prison, still protesting his innocence.

MURDER BY INSULIN

Since its discovery and widespread use as a treatment for diabetes, insulin has received an undeserved reputation as an efficient and undetectable method of murder. In fact, it is neither efficient nor undetectable. Administering enough insulin to induce hypoglycemia and ultimately death can take a while; the symptoms of hypoglycemia are easily diagnosed; and the cure is nothing but simple glucose. Although commercial insulin acts in the same way as the body's natural insulin, small tweaks in the sequence of amino acids making up pharmaceutical insulin (to make it faster- or slower-acting) allow for easy confirmation that any excess insulin in a murder victim got there by nefarious means. In fact, murder by insulin poisoning is quite rare, with fewer than seventy documented cases worldwide, the vast majority occurring in the UK and the United States. It is quite disheartening to know that most have been committed by doctors, nurses, and other health care professionals.

Traditionally, patients with diabetes have relied on multiple daily measurements of glucose in their blood from a finger prick to determine the correct amount of insulin to self-inject. An available alternative is an insulin pump, a computerized device about the size of a smartphone, which delivers insulin through a catheter placed into the fat layer under the skin.[9] Amazingly, some insulin pumps also have the capability of continuously monitoring blood glucose levels to provide real-time information on the correct amount of insulin to be infused, essentially taking on the role of the pancreas.

Since the pumps are run by small computers, there is a risk that cybersecurity vulnerabilities in the software could be hijacked to cause a lethal dose of insulin to be injected. Could someone with a grudge against the pump manufacturer, or resentment toward an individual with diabetes, really commit murder over the internet? In 2019 a major insulin pump manufacturer recalled some of its pumps due to a glitch that could be exploited by nearby hackers to take over control of the pumps.

Being the first person to do something is usually a cause for celebration. In Kenneth Barlow's case, his desire to be the first person to commit murder with insulin was thoroughly dishonorable. The prosecution's witnesses confirmed that Kenneth had spent a lot of time thinking about insulin as the perfect murder weapon. No doubt Kenneth assumed that since no one had been murdered with insulin before, its novelty would cause it to be overlooked should any questions arise concerning his wife's death. Unfortunately for him, Kenneth's innovative way of committing murder did little to help him evade the consequences of his actions.

In the next chapter we move from a drug that had never been used as a poison before to one that has a long and storied history of being used not only as a method of murder but also as a cosmetic no self-respecting Renaissance lady would be without.

. { 2 } .

Atropine and Alexandra's Tonic

"My goodness," he cried, "I've only just realized it! That rascal, with his poisonous cocktail!"

—Agatha Christie, *Three Act Tragedy*, 1934

DRUG PLANTS

Members of the Solanaceae plant family, including such familiar foods as potatoes, eggplants (aubergines), chili peppers, and tomatoes, are eaten daily in homes across the country. Yet despite their regular appearance on present-day dinner tables, Solanaceae plants were initially treated with concern and suspicion. Tomatoes, brought to Europe from the New World by sixteenth-century Spanish conquistadores, saw merchants struggling to peddle this new fruit to consumers, who were convinced it would kill them. To combat this tomato phobia, vendors often employed people to eat tomatoes next to their stalls, giving what amounted to some of the earliest consumer reviews of products. The widespread use of tomatoes in cooking today is a clear testament to those early daring customers.[1] But why were so many people deathly afraid of the humble tomato?

The answer lies in other members of the Solanaceae family, which may look like potato and tomato plants, but ingestion of which can be deadly. Among these is *Atropa belladonna*, an attractive plant with violet flowers that yield small shiny berries displaying a deep purple or black hue. A single berry incorporated into food or drink contains enough poison to cause death. Although the plant's Latin name hints at its toxicity, its common name, deadly nightshade, leaves no doubt regarding its lethal qualities.

In Greek mythology, when a newborn was three days old, three Fates would visit it to determine its destiny. Clotho (the Spinner), the youngest of the Fates, spun the thread of life from dark and light fibers. Lachesis (the Apportioner) determined the length of the thread of life, and Atropos (the Inevitable) held the shears to cut the thread of life, sealing the length of the person's life. Little wonder, then, that Atropos lent her name to belladonna's most lethal component, atropine.[2]

Atropine, when pure, is a white odorless crystalline powder, first purified from *Atropa belladonna* berries and leaves in 1833 by the German chemist Philipp Lounz Geiger and his Swiss student Germain Henri Hess.[3] Atropine is chemically classed as a plant alkaloid, sharing similarities with other alkaloids. These compounds, when dissolved in water, typically produce alkaline solutions. They also tend to be very bitter tasting. Although the small shiny berries may look tempting to eat, anyone who unwittingly chews a berry will immediately spit it out due to its bitter taste. For this reason very few people die from accidental atropine poisoning.

The "belladonna" part of *Atropa belladonna* comes from the Italian *bella donna*, meaning "beautiful lady." In 1544 Pietro Andrea Mattioli, an Italian physician and botanist, published *Materia Medica*, in which he described his concept of medicinal botany. Although Mattioli was a physician and healer, he also performed studies on the use of poisonous plants in daily life. Among his observations were that Venetian actresses and courtesans would squeeze a drop of juice from a belladonna berry into their eyes to dilate their pupils and achieve a fashionably alluring appearance. It is alleged that part of the attraction of Da Vinci's *Mona Lisa* is due to the belladonna-induced dilation of the Mona Lisa's eyes. Developing an attractive doe-eyed appearance was not without its drawbacks, however. Dilation of the pupil should permit more light into the eye to see clearly, but the atropine in belladonna also has the effect of relaxing the muscles controlling the lens, so that courtesans really couldn't see who they were flirting with. Prolonged use of belladonna is also likely to lead to blindness. Although belladonna berries are no longer used to dilate pupils for romance, the same premise is still employed by restaurants that use dim lights and candles on the tables, causing the pupils to dilate spontaneously and allow more light into the eye.

The opposite effect occurs when we step out into bright sunlight and the pupils rapidly contract to prevent damage to the retinas. Such rapid changes in the pupils in response to varying light intensities is due to the actions of nerves on the small muscles controlling pupil size. The observation that belladonna juice (atropine) also affects pupil size implies that it is somehow interfering with the normal transmission of information from nerves to muscles.

To understand exactly how atropine not only affects pupils but can lead to death, we need to take a little detour into a scientific argument that was raging across Europe in the late 1800s.

SOUPS AND SPARKS

How does the brain tell a pupil to dilate or contract, an arm to move, a hand to turn the pages of this book, or the heart to beat faster or slower? This seemingly straightforward question would, in the waning years of the nineteenth century, become one of the most acrimonious debates in biology. Lined up on each side of the argument, like soldiers set to charge at each other across a battle line, were distinguished scientists, each committed to the supremacy of their own ideas and convinced that anyone who opposed them was willfully ignorant.

At the end of the nineteenth century the reticular theory of the nervous system, including the brain, stated that it was made from a single, large, continuous network. The Nobel Prize winner Camillo Golgi put his considerable academic prowess behind this concept, making it the prevailing scientific theory of nerves at the time. That is, until the Spanish scientist Santiago Ramón y Cajal, came along and told everyone who would listen that the reticular theory was complete nonsense.

Based on careful examination of hundreds of brain slices,

Cajal proposed his neuron doctrine, in which he concluded that the nervous system was not one gigantic mesh, but was in fact composed of many individual nerve cells, and that between each nerve cell was a tiny gap called a synapse. To get some idea of how small this gap is, in the metric system one inch is about 25.4 million nanometers (a nanometer is 1,000,000,000th of a meter). A human hair is about 80–100,000 nanometers thick, a sheet of paper is approximately 100,000 nanometers, and a synapse is between 20 and 40 nanometers. Although this gap is incredibly small, it is still nonetheless a gap.

The primary question at the turn of the twentieth century was how information crossed that gap. Always up for a good argument, scientists divided themselves into two camps. There were those who believed pulses of chemicals were sent across the gap—adherents of this position calling themselves "soups"—whereas those who believed that a jolt of electricity spanned the gap were self-styled "sparks." Not even political meetings can attain the acrimony and bitterness of two groups of scientists, each convinced of the merits of their position, and the lack of any credibility in their opponents' platform. Indeed, this scientific feud would shape neuroscience for the next fifty years.

For much of the nineteenth century, scientific theory had been dominated by German chemists, but the science of electricity was starting to take hold. In 1791 Luigi Galvani demonstrated that frogs' legs could be made to twitch when stimulated with electricity. Indeed, such early experiments with electricity on animal tissues became one of the strong influences on the young Mary

Shelley as she wrote *Frankenstein* in 1818. As the twentieth century dawned, electricity seemed new, modern, and exciting, whereas chemicals were so last-century. The concept of electrical signals carrying information across gaps reached its peak with the work of Guglielmo Marconi, who, in 1901, showed the world how to communicate using wireless radio transmissions. If the electromagnetic waves of wireless radio could travel hundreds of miles, surely they could cross the tiny gap at the synapse.[4]

More important, there was evidence in favor of the "sparks" position. The ability to make fine electrical wires had just been developed, and when these were inserted into nerve cells, it was found that the nerves always had an electrical discharge when they fired. To be fair, this was seen only inside a nerve cell, but it wasn't much of a stretch to imagine that the electrical discharge could also occur across the tiny gap of a synapse. Further support for the "sparks" theory came from experiments on frogs' hearts. It was known that if you took out the heart from a frog and placed it in a beaker of salt water, or saline, it would continue to beat as if it were still in the frog. If the dissection was done carefully, some of the nerves attached to the heart could be kept intact. Hooking up electrodes to a battery, scientists could stimulate various nerves attached to the heart, causing it either to slow down or speed up. Surely this was evidence that the "sparks" were right.

Not to be outmaneuvered, the "soups" got their own beakers of saline and frogs' hearts. Instead of batteries and wires, the "soups" added various chemicals to the beaker, and they, too, found that it was possible to cause a heart to speed up or slow down, depending on which chemicals were added to the liquid. However, the "sparks" were quick to point out that all these chemicals were human-made, coming from chemists' workbenches, and therefore more sideshow than biology.

Intrigued by the sparks-and-soups debate, a young German scientist, Otto von Loewi, brashly decided that he would solve the riddle. If you were to look up "absentminded professor" on the internet, you might find a picture of Loewi. Even as a student, he often neglected his biology classes, favoring trips to the opera or philosophy lectures instead.

Easter 1920 would prove a pivotal point for Loewi and the emerging science of neuropharmacology (the study of how drugs affect nerves, and especially nerves in the brain). The Saturday night before Easter, Loewi was at home reading. The book he was reading was apparently so engrossing that he promptly nodded off to sleep. While asleep, Loewi had a dream in which he performed experiments that solved the "soups" versus "sparks" dilemma once and for all.[5] In a half-groggy state, he scribbled notes on a tiny slip of paper as to how he should perform these groundbreaking experiments. Exhausted from his dreams and frantic note-taking, Loewi fell back to sleep. Awaking at six the next morning, he recalled that he had written down something important the previous night—but was devastated to find that he was unable to read his own scrawl. The next day was spent trying to tease out any sense from his nocturnal scribbles, but to no avail. Crushed that a momentous opportunity had seemingly passed him by, Loewi retired to bed for the night.

Incredibly, early the next morning, the dream returned. This time, not trusting his dream to illegible jottings, Loewi leaped out of bed and went immediately to his laboratory. There he euthanized two frogs and removed their hearts, placing them in two separate beakers of saline, and watched them beat, as he and others had done so many times before. Using wires to electrically stimulate the vagus nerve of the first heart caused the heart to slow down—just as he expected. The next step involved doing something no one had ever thought to do before. With trembling hands, Loewi used a dropper to suck up the saline that had been bathing the first heart, and dripped it over the heart in the second beaker. To his delight, Loewi observed that the second heart also slowed down, despite never being exposed to any electrical input.

Excitedly Loewi returned to the first heart, where stimulating a different nerve caused the heart rate to speed up. This time, pouring the saline from the first heart over the second heart caused the second also to speed up—exactly as his dream had predicted. Loewi concluded that electrically stimulating the vagus nerve from the first heart had released some chemical into the saline that caused the heart to slow down, a chemical that entered the saline and could be transferred to a second heart to make it, too, slow down. Not one to be eloquent, Loewi called this chemical stuff released by the vagus nerve *Vagusstoff*. (German for vagus matter.) We now know *Vagusstoff* to be the neurotransmitter acetylcholine. The experiments arising from Loewi's dreams were to bring him a Nobel Prize in Physiology in 1936.

Did Loewi solve the riddle of the soups and sparks, and prove who was right? Well, yes—and no, since the answer is both! We now know that when a nerve "fires" and the signal ripples down its entire length, that's the electrical component that the "sparks" detected. But when the electrical signal comes to the end of the nerve, electricity cannot jump across the synapse, and so the nerve converts its electrical message into a chemical one. Like a chemical supply store, the nerve ending stores chemical messages, or neurotransmitters, in little packets, all ready to be released into the synapse when given the appropriate indication. Depending on the message needing to be communicated, there are different kinds of neurotransmitters. When triggered, the packets of neurotransmitter get dumped into the synapse, allowing the chemicals to cascade across the gap and dock with special docking proteins, or receptors, on adjacent cells. This is the chemical component the "soups" detected.

What happens next depends on what is on the receiving side of the synapse. It may, for example, be a sweat gland, with reception of the chemical message increasing sweat production, or it may be a pancreas being stimulated to release digestive enzymes into the intestine. Not all signals are received as messages to increase activity. Some chemical signals, like Loewi's acetylcholine, can be received by the heart as signals to slow down. As we will discover shortly, it is the reception of the chemical message on the other side of the synapse that is affected by atropine, essentially shutting down the signaling process entirely and disrupting the brain's normal control of the body.

The bitterness of atropine makes accidental poisoning through eating a belladonna berry almost impossible. For would-be murderers, the trick is disguising atropine's bitter taste. The solution to that difficulty lies in a problem confronting the British army in India during Queen Victoria's reign.

GIN AND A DEADLY PLAN

In nineteenth-century colonial India, British army officers and enlisted men were being bitten by mosquitoes and coming down with malaria. So big was the problem that in the mid-1800s, life expectancy in the East was half of what it was back in Britain. With so many of the soldiers and government workers bedridden with malaria, governance of the subcontinent was becoming extremely difficult. The solution came from studies by the Scottish physician George Cleghorn, who discovered that the bark of the cinchona tree contains a compound—later identified as quinine—which when dissolved in water and imbibed, proved very effective against malaria. Although the quinine tonic was helpful in staving off the disease, its bitter taste made it quite unpleasant to drink. British officers took to adding sugar, lime, and gin to the tonic to make it more palatable, and the gin-and-tonic cocktail was born—purely for medicinal purposes.

Some 150 years later, the drink designed to mask the taste of a bitter drug was set to cause panic across the city of Edinburgh, Scotland.

MAYHEM AT THE MARKET

At the end of August 1994, a constant stream of calls to the police hotline was overwhelming officers in Edinburgh. Most of the calls would not provide anything useful, but they still had to be answered and meticulously logged. Detective Superintendent John McGowan paced through the incident room. Contaminated items had been found on grocery store shelves, and people were becoming ill. It was still early in the case, but he had no clear indication of a motive. Was it terrorism, blackmail, or an indignant ex-employee? It was hoped that figuring out why the crime was committed would provide a clue as to who was behind the seemingly random attacks. Was it an accident or was it deliberate?

A few days earlier John and Marie Mason had gone to do their weekly food shopping at a local Safeway grocery store in Hunter's Tryst, in the Edinburgh suburbs. Once home, the couple unpacked the groceries and started putting things away, when Marie noticed that they had forgotten to buy any tonic water. Hardly a major catastrophe, but Marie liked to keep some on hand to help settle her stomach when she got a little heartburn. John, an appropriately dutiful husband, took the trip back to the store to pick up some tonic—a decision that would dramatically affect the Masons' lives. Upon arriving home, John poured his wife a glass of tonic water. A short while later, she said that she felt unwell and would go to bed early. As she undressed for bed, Marie stumbled and fell over, an unusual occurrence for her, but she figured she was just tired. The next day she awoke still feeling unwell; to soothe her aching stomach, she drank two more glasses of tonic, little knowing it would make her condition worse. Marie's vision got blurry, and she began to hallucinate, telling her husband that water was pouring out of the radiators. Marie was taken to the Royal Infirmary hospital, where doctors tried to determine what was wrong.

Unknown to the Masons at the time, Elizabeth Sharwood-Smith had also bought some tonic from the same grocery store. That weekend Elizabeth and her eighteen-year-old son, Andrew, complained of terrible stomach cramps and sickness, to the point where both were taken to the emergency room. Over the course of that fateful weekend, four people would be rushed to hospital after drinking contaminated tonic water, and in total, eight people would fall victim to poisoning.

Given the scale of the poisoning, Detective McGowan was tasked with setting up an incident room, and the calls were flooding in, most of them with conflicting and irrelevant information and not a few time wasters. A press conference was called by Safeway, and anyone from Edinburgh who had purchased tonic from the Hunter's Tryst store was asked to return it. Of those bottles that were returned, six more were found to be contaminated. Although the tainted bottles were found only in Edinburgh, a nationwide panic was set off and fifty thousand bottles of Safeway tonic were removed from the shelves and destroyed. Speculation as to the cause of the contamination ran rampant in the press. Was a poisoner loose on the streets of Edinburgh? Had there been a terrible contamination at the bottling plant?

In fact the tonic Marie, Elizabeth, Andrew, and others drank was all part of an elaborate smokescreen to hide the poisoner's true intent: a plot to kill his wife and live a new life with his mistress.

Born in Glossop, in the foothills of the Derbyshire moors, Paul Agutter showed himself to be highly intelligent, and because of strong exam results at school, earned a coveted place at Edinburgh University. Agutter studied Biochemistry and obtained a first-class bachelor's degree in the subject in 1968. Well known to the science faculty, Agutter remained at the university, where he worked as a demonstrator in undergraduate biochemistry practical labs while earning his Ph.D. in Molecular Biology. Upon graduating with his doctorate, Agutter took up an academic position as a lecturer in Cell Biology in the School of Life Sciences at Napier University, on the southern outskirts of Edinburgh. As often happens, Paul married another academic, Dr. Alexandra Agutter, who taught university-level English.

To all outward appearances, the Agutters enjoyed an enviable married life, often having friends over for dinner at their home in Kilduff Lodge, in the historic town of Athelstaneford in East Lothian, some twenty miles east of Edinburgh. Inside the marriage, though, things were rather different. Paul complained of depression to his doctor, a depression so deep that he contemplated taking his own life. Financial and marital troubles further compounded his struggles.

Perhaps Paul was just feeling the pangs of a midlife crisis, but the light at the end of the tunnel, at least from his perspective, was one of his students at Napier University, a rather attractive woman called Carole Bonsall. In Paul's mind marrying this student, who no doubt stroked his ego and declared him to be the finest mind in the university, would solve all his problems. However, before he could achieve this life of paradisiacal bliss, a major barrier had to be overcome: He was already married. Divorce would mean being ejected from his house, and his precarious finances would be thrown into further turmoil. Life would be so much easier for Paul if only his wife were dead. Since Alexandra did not appear to be particularly cooperative in dying, Paul decided to hasten her demise by plotting her murder.

As a lecturer in Biology, Paul had some knowledge of poisons; he also knew that many of these poisons were easily detected at postmortem. Despite this, Paul was confident in his intellect, and, as an avid chess player, was used to planning things several steps ahead. Working in a toxicology research group at Napier University, he had easy access to atropine. Since atropine is an easily detected poison, the finger of suspicion should not point at Paul, but perhaps to some imaginary mass murderer loose in the city.

There are two key elements to committing the perfect murder: naturally, the intended victim should die; but the murderer should also escape arrest, conviction, and imprisonment. Despite his cunning schemes, Paul Agutter was destined to fail in both these respects. Paul bought several bottles of store-brand tonic, which he deliberately spiked with atropine taken from his research lab. On Wednesday, August 24, 1994, he placed his doctored tonic bottles back onto the shelves of the Safeway grocery store at Hunter's Tryst, a short drive from his office and laboratory. The bottles all had some atropine in them, not enough to kill, but enough to make anyone who drank them very ill. Paul kept one of the bottles for himself, waiting to add more atropine to make a lethal beverage. In this way he envisioned that the atropine-induced death of his wife, Alexandra, would be seen as only part of a larger scheme by some unknown person or persons to cause mass panic among Edinburgh's populace. In this respect he was initially successful, as people like the Masons and the Sharwood-Smiths would later go into Safeway and purchase the tainted tonic, leading to outbreaks of illness across the city.

Paul knew that one drawback to his plan was that atropine tasted extremely bitter, so he had decided the best solution was to mask it in something else—just as British officers had done with bitter quinine in India more than a century before. As his wife relaxed on the warm summer evening of August 28, a chilled gin and tonic appeared to be the perfect beverage. Paul poured his wife a large drink, splashed with atropine-laced tonic, and waited for it to take effect. Alexandra sipped the drink, and then sipped again; something didn't taste right. The drink was a little too bitter, and instead of drinking the whole glass, Alexandra drank only part of it. Even so, it could have been enough to kill her. It certainly was enough to bring on all the symptoms characteristic of atropine poisoning. Her mouth went dry, her heart started racing, and when she stood up, dizziness caused her to collapse on the floor. At that point the hallucinations began, with Alexandra later recalling that everything looked as if it were made of gossamer silk.

Seeing his wife's distress, Paul Agutter calmly announced he would call for help. Rather than an ambulance, however, he called their local primary-care physician, who happened, as Paul had confirmed earlier, to be out of town. That was good news for

Paul; the last thing he needed was a doctor coming around to treat his wife. To further strengthen his alibi, Paul left an urgent-sounding message on the answering machine, requesting the doctor to visit as soon as possible.

This is where Paul's carefully crafted plan began to unwind. Unexpectedly, his message was picked up by a stand-in doctor who was on call that night, and when the doctor got to the Agutters' home, he immediately saw that Alexandra was very ill, possibly poisoned by something she had eaten or drunk. The doctor called for an ambulance to take Alexandra to the hospital, and when the ambulance staff arrived, they asked what Mrs. Agutter had last had to eat or drink. She pointed to the half-drunk glass of gin and tonic on the small table next to her chair. The paramedics took not only the glass but also the (spiked) bottle of Safeway tonic. The fact that Mrs. Agutter didn't drink the whole glass of gin and tonic likely saved her life, although she remained seriously ill for some time.

By the end of the weekend, eight people had been taken to the hospital and diagnosed with atropine poisoning. What was the common thread between these seemingly unrelated poison victims and Mrs. Agutter? The conclusion was that all the victims had obtained tonic from the same Safeway. The working theory of the police was that some crazed individual had laced the tonic with poison to blackmail the supermarket. Keeping up his pretense, Paul Agutter was interviewed by the press on his reaction to his wife's poisoning. Calmly, Paul stated that he could not comprehend how someone could do this, lamenting, "My wife could have been killed" and adding that "the attempted murder of my wife [and others] is outside anything I can understand." He begged the culprit to come forward and surrender to the police as soon as possible—all the while knowing that he himself was the man police were looking for.

Much to Paul's delight, twenty-six-year-old Wayne Smith wrote to a local newspaper claiming to be the poisoner. However, this respite for Paul was short-lived, as the police tracked the man down and brought him in for questioning. Since Smith didn't know any of the fine details of the case, such as how many bottles had been tainted, the police quickly realized that the man had nothing to do with the poisoning.

Things started to look bad for Paul Agutter when forensic scientists measured the amounts of atropine in all the collected bottles. Most contained between 11 and 74 mg of atropine, but Agutter's bottle stood out at a whopping 300 mg of atropine. Too late, Paul realized he should have removed his wife's gin and tonic and his deadly tonic bottle, or replaced it with one containing a similar level of atropine as found in the other bottles, before the ambulance crew whisked it away. Had he changed the bottles, it is unlikely he would ever have been suspected of his wife's attempted murder. It was later determined that Alexandra had probably ingested only 50 mg of atropine because she didn't finish her drink due to its bitter taste.

As the net closed around Agutter, closed-circuit TV camera footage revealed him in the Safeway a few days before people succumbed to the poisoning; unfortunately the camera didn't show Paul actually putting bottles back on the shelf. Remarkably, though, a Napier student who knew Paul was working that day as a stock boy and had in fact seen him putting several bottles back on the shelf. When confronted with this evidence by the police, Paul brazenly stated that of course he had handled tonic bottles at the store, since that was where he had purchased the tonic for his wife. However, the fact that the Agutter's tonic bottle contained significantly more atropine than any of the other bottles argued against him.

In 1995 Paul Agutter was finally arrested and sent to trial for the attempted murder of his wife. Throughout the trial, one of his staunchest defenders was his wife, Alexandra, who believed her husband incapable of murder. Nonetheless, Paul Agutter was found guilty of attempted murder, and at his sentencing the judge said: "This was an evil and cunningly devised crime which was not only designed to kill your wife, but also to cause great alarm, danger, and injury to the public." Paul was sentenced to twelve years in prison.

The postscript to Paul's attempt at murder is equally remarkable and highlights the observation that fact is often stranger than fiction. During his incarceration, Paul Agutter was assigned an interesting cellmate—none other than Wayne Smith, the man who had made the false confession of poisoning the tonic. After his confession was exposed as fraudulent, Smith decided to try the real thing and was found guilty of putting weed killer in cartons of fruit juice at another Safeway.

In prison Agutter spent time working in the library helping other inmates learn to read. Alexandra Agutter finally accepted that her husband had tried to kill her, and divorced him while he was in prison. Carole Bonsall, the mistress for whom he was willing to murder his wife, ditched him and wanted nothing more to do with him. In 2002, a sad and lonely fifty-eight-year-old Agutter was released from prison after serving seven years of his twelve-year sentence. The parolee left Scotland and moved back to Derbyshire to live with his aging parents. Remarkably, he got a job for a time at Manchester University lecturing in evening classes; the course he taught? Philosophy and Medical Ethics.

HOW ATROPINE KILLS

Atropine works on the part of the nervous system called the parasympathetic nervous system. This is the part of the nervous system that enables the body to "rest and digest." The better-known sympathetic nervous system, in contrast, is associated with the fight-or-flight response. Not surprisingly, these two systems rely on different chemical transmitters to cross the gap of the synapse. In the case of the parasympathetic system, the neurochemical involved is acetylcholine. When we sit down to eat a meal, the parasympathetic nervous system stimulates the production of saliva in the mouth; the typical "mouthwatering" sensation we get when we smell a particularly delicious meal being cooked. Farther down in our intestines, the same nervous system tells the pancreas to release more digestive enzymes to break down our food. When we are relaxed and comfortable, acetylcholine encourages our heart rate to become lower as we enter a state of contentment.

Acetylcholine works because it has just the right shape to fit into its receptor on the far side of the synapse, just as the correct key will easily fit into its proper lock. As such, acetylcholine is referred to as an agonist. Although only the right key will fit into the lock and unlock it, other keys may have a similar-enough shape that they will go into the lock but are unable to unlock the mechanism. Annoyingly, the wrong key can often get stuck in the lock, preventing the right key from being used. Atropine is such a wrong key: It looks like acetylcholine and binds to the receptor but doesn't activate it, and also prevents any acetylcholine from entering and activating the receptor. Physiologically atropine is acting as an antagonist. In the presence of atropine the normal signals that acetylcholine conveys never get through, and whatever effect acetylcholine would cause, the opposite occurs.

The parasympathetic nervous system, through acetylcholine, stimulates saliva production; when this is blocked by atropine, the mouth becomes "dry as a bone." An excessively dry mouth also results in severe thirst, together with difficulty in swallowing. Tears can also dry up, making the eyes itchy and red.

Acetylcholine helps contract the pupils so we can more easily focus on what's right in front of us, rather than use peripheral vision to scan for potential danger. Atropine prevents that contraction causing the dilated pupils so favored by Italian courtesans. It also causes relaxation of the muscles controlling focus of the eye, so that victims are wide-eyed and blind, effectively "blind as a bat."

Usually when we digest a meal, blood is diverted away from the skin toward the intestines, so that absorbed nutrients can be carried around the body. When the effects of acetylcholine are blocked by atropine, the blood vessels of the skin open up, giving a ruddy appearance to the complexion, so that victims appear "red as a beet."

Atropine can also affect nerves in the brain, leading to rambling, incoherent speech, an inability to walk in a straight line, and eventually hallucinations, generally resembling a person who is drunk or "mad as a hatter." Hallucinations associated with atropine poisoning are visual and very realistic, with commonly reported examples of butterflies, trees, faces, snakes, or even silk curtains; all this in contrast to the psychedelic hallucinations induced by drugs such as LSD (lysergic acid diethylamide).

Finally, atropine poisoning also affects the body's ability to control its temperature, causing victims to be "hot as a hare."

Acetylcholine causes the heart to slow down (remember Loewi's experiments), but atropine blocks the actions of acetylcholine, and so the heart receives no "slow down" signal. Instead, the heart rate slowly accelerates, ultimately leading to a very rapid heart rate of 120 to 160 beats per minute. Not only does the heartbeat become very fast but it also starts to become irregular and can even stop altogether, leading to heart failure and death. The fast heartbeat caused by atropine also leads to severely elevated blood pressure, which can cause problems for the kidneys and the brain.

How quickly atropine acts on the body depends on how it enters. When injected directly into the bloodstream, atropine's effects can be felt within a few minutes, but if it is present in food or drink, the effects may be noticed within a quarter of an hour. Within the body atropine has a half-life of about two hours; that is, half of the atropine in the body can be eliminated within two hours. About 50 percent is directly filtered by the kidneys and excreted in urine, and the rest broken down by enzymes in the liver. Even so, it can take several days for all traces of atropine to be removed, and hallucinations can continue for many hours.

Another plant in the Solanaceae family containing atropine is *Datura stramonium*, also known as devil's snare or jimson weed, after the town of Jamestown, Virginia. The colonial governor of Virginia sent troops to quell a rebellion in 1676, and while some of the soldiers waited for reinforcements, they collected the leaves off a local plant, to boil and add to their meal. The hallucinogenic properties of the plant took effect immediately. Soldiers would sit naked in corners, grinning like monkeys and pawing at passersby. Others would blow feathers in the air, or take great interest in straw stems. The soldiers were eventually confined and treated, but it took eleven days for them to fully recover from their ordeal.

DR. BUCHANAN, THE MADAM, AND THE DEAD CAT

One of the frequently recurring themes in this book is the idea that physicians and scientists believe their training and expertise provide unparalleled insights that allow them to commit undetectable murders where others had failed. On May 8, 1893, Dr. Carlyle Harris was electrocuted at Sing Sing prison in New York, for the murder of his wife. Harris's poison of choice was morphine, overdoses of which depress brain activity, eventually causing breathing to stop. Death by morphine overdose can easily be confused with death by natural causes (hence its use by numerous murderers), but it does leave one telltale sign of its presence. Morphine causes the pupils to contract markedly, leading to the classic characteristic of morphine overdose: pinpoint pupils. It was partly because the coroners noted contracted pupils on the corpse of Harris's wife that he was arrested and tried for murder by morphine poisoning.

Shortly after Harris's execution, another New York physician, Dr. Robert Buchanan, was convinced Harris had been caught because he was inept. In fact, Buchanan acquired a habit of sitting in bars, drinking heavily, and announcing to anyone who cared to listen (and often those who didn't) that Harris was a bungling incompetent. Buchanan opined that it was eminently possible to disguise morphine poisoning, simply by using a drug that would dilate the pupils, thus eradicating the most obvious sign of morphine overdose: atropine.

Robert Buchanan was born in Nova Scotia in 1862. In 1886 he, along with his wife and daughter, moved to New York to start his new medical practice. Coming from a Canadian province with a total population of around 31,000 to a city of 1,500,000 people must have been a culture shock, but Buchanan took full advantage of the big city. Outwardly respectable, Buchanan was far from respectable outside of his professional life, with a liking for heavy drinking, and a definite fondness for brothels. One of the brothels was owned by Anna Sutherland, with whom Buchanan started an affair. Fortunately for Buchanan, his wife, friends, acquaintances, and—more important—patients had no idea about his extracurricular activities.

However, as is true of so many affairs, the truth soon came to light. At first Buchanan said that Anna Sutherland was merely one of his patients, and he was seeing her out of professional courtesy. Buchanan's wife didn't buy that for a second, and in the summer of 1890 the couple were divorced. Considering the events that were about to transpire, she was lucky to escape with a divorce.

Although 1890s New York was rough and tumble, the clientele Buchanan sought for his medical practice were from the more genteel sector of society, and they didn't take too kindly to their doctor installing a known brothel owner as his new receptionist. Anna Sutherland, some twenty years Buchanan's senior, became besotted by her new suitor, changing her will to make Buchanan the sole beneficiary of her estate. Always forward thinking, Buchanan also persuaded Sutherland to take out a five-hundred-thousand-dollar life insurance policy, again with Buchanan as the sole beneficiary.

Distressingly to Buchanan's patients, not only was his receptionist a brothel madam but she was also vulgar and crude. Patients began switching to other doctors, and even those who stayed were only in the process of finding a new physician. With a dwindling income stream, Buchanan's expensive lifestyle soon ate into his bank account. It seemed that Anna Sutherland was becoming a liability, and Buchanan, impressed by his own intelligence, knew exactly how to take care of the problem.

On Friday morning, April 22, 1892, after eating a hearty

breakfast, Anna became very ill, feeling severe pains in her stomach and unable to stand. Taking to her bed, she was visited by Dr. McIntyre, a physician acquaintance, who found her in excruciating pain, complaining of headaches, and having difficulty breathing. Displaying all the compassion and empathy he could muster, McIntyre diagnosed hysteria and prescribed a small sedative. Notwithstanding such care, Anna was still unwell by the afternoon, at which time Buchanan was seen giving her a few teaspoonfuls of medicine, medicine his wife complained tasted almost too bitter to take.

At seven o'clock that evening, Dr. McIntyre returned to see how his patient was doing. By then Anna was in a deep coma, with a rapid pulse, very shallow breathing, and her skin was hot and dry. It wasn't long before Anna was dead, likely because of "cerebral apoplexy," or cerebral hemorrhage.

Buchanan's problems also now seemed to be gone, as he inherited Sutherland's money and property and was the recipient of the large life insurance policy. With a large bank balance and the loss of a vulgar receptionist, Buchanan's life seemed to be on the upswing, so much so that his first wife (clearly more forgiving than most other wives would be in the same situation) agreed to remarry him.

But things began to unravel slowly. Buchanan's drunken braggadocio and contempt for Carlyle Harris would come back to haunt him. A journalist looking for barroom gossip heard about Buchanan's ramblings, and was suspicious enough to start digging. The journalist learned of Buchanan's severe financial problems before Sutherland's untimely death, and that he was the sole beneficiary of her estate. Contacting the police, the reporter relayed all that he had found, generating enough suspicion to warrant an exhumation of Sutherland's body. An examination of the liver and intestines showed the presence of morphine, in an amount consistent with causing death.

How was the level of morphine in Anna Sutherland's tissues determined? The reader may recall how insulin, extracted from Elizabeth Barlow's buttocks, was injected into mice to determine its level; a similar method was used to quantify morphine, although in this case extracts were injected into frogs to determine how much of the extract was needed to kill the frog. Despite the finding of lethal morphine levels in Sutherland's body, one key element of morphine poisoning—contracted pinpoint pupils—was absent. Had Sutherland really just died from cerebral hemorrhage, or had Buchanan found a way to hide the telltale sign of morphine overdose, as he had bragged?

Whatever the merits of the case, Buchanan was arrested and charged with first-degree murder. The trial was a sensation, and the first in America in which forensic evidence was brought in by the prosecution. The defense argued that there was no evidence that Anna had died from a morphine overdose, as her pupils were wide and dilated. The prosecution, in turn, brought a stray cat into the courtroom, and before a jury, at once enthralled and disgusted, killed the poor cat with a lethal dose of morphine. (Exactly what the cat thought of these courtroom theatrics was not recorded.) As the dead cat's eyelids were pulled back, the characteristic pinpoint pupils of a morphine overdose were clearly visible. Then atropine was slowly dripped into the animal's eyes. As the jury watched in grim fascination, the cat's pupils slowly but inexorably dilated until they were fully open.

Buchanan's barroom theory that he could use atropine to disguise the effects of morphine poisoning had been proved correct; unfortunately for him, it was also proved correct to the jury. On April 25,1893, the jury brought back a verdict of guilty, with no recommendation for mercy. With no other options open to him, the judge gave Buchanan a mandatory death sentence.

Although Buchanan had called Carlyle a bungling incompetent, Buchanan's final days were remarkably similar to Carlyle's. He was transported to Sing Sing Prison under heavy guard to await his sentence. When all his appeals were exhausted, Buchanan slowly realized that, in the end, he had been no smarter than Carlyle. Walking the twenty yards from his cell to the electric chair, Buchanan remained impassive and silent. As the electrodes and straps were applied to Buchanan's body, the state electrician was given the signal to close the switch. Two minutes later Buchanan was dead.

ATTEMPTED ASSASSINATION OF THE SALISBURY SOVIET SPY

Although Paul Agutter attempted to kill his wife with atropine, and Dr. Buchanan used atropine to avoid detection in the murder of his wife, what is quite remarkable is that atropine is also the remedy for far deadlier nerve poisons. These lethal agents can exist either as a liquid, "painted" on door handles or other solid surfaces to be absorbed through the skin, or they can be made into gases and absorbed through the lungs. No matter how the nerve agent enters the body, once it's inside, it causes damage in the exact same way. We've examined the problems that occur when too little acetylcholine reaches its receptor, but overstimulation with too much acetylcholine can be just as lethal.

Once acetylcholine is released from the nerve ending and crosses the synapse to bind and activate its receptor, the acetylcholine must be broken down quickly to prevent the signal from becoming overwhelming. The job of breaking down acetylcholine is performed by an enzyme, appropriately called acetylcholinesterase, taking just 80 microseconds (80 millionths of a second) to degrade each molecule of acetylcholine. Nerve agents attack the acetylcholinesterase enzymes, putting them out of action and destroying their ability to breakdown acetylcholine. Meanwhile, acetylcholine continues to be released from nerve endings, leading to a massive accumulation of acetylcholine, triggering the receptors over and over until the target organ finally malfunctions.

What effect would this have on a victim exposed to a nerve agent? We saw, for example, that a small amount of acetylcholine helps slow down heart rates when we are resting; with considerable amounts of acetylcholine continually bombarding the receptors, heart rates can plummet to dangerously low levels. Excess acetylcholine will cause excessive production of sweat, tears, and saliva, so much saliva that it can appear as if the victim is frothing at the mouth. Sweating can become so profuse that the victim's clothing will often be drenched. The normal low level of fluid secretion that keeps the lungs and airways moist and clean goes into overdrive, to the point that victims can start drowning in their own secretions. Headaches, with subsequent convulsions, unconsciousness, and coma, as well as nausea and vomiting, will accompany all these symptoms. Unless treatment is started soon after exposure, victims of nerve agents are unlikely to survive. The only treatment for nerve agents is an unlikely drug, one which on its own can be deadly, but when used to treat victims exposed to nerve agents is remarkably efficacious: atropine.

It would seem surprising that a deadly chemical such as atropine could ever be an antidote that offers protection against another poison. In fact, it is only recently that atropine has become the treatment for a group of chemicals that were invented in the 1940s—the organophosphates. Originally developed as pesticides, organophosphate compounds have been further developed to become some of the deadliest chemicals ever made, and include VX and VR nerve gases, as well as sarin and Novichok.

Sergei Skripal was a colonel in the GRU (Russian Military Intelligence). During a posting in Madrid, Spain, Skripal was approached and recruited by the British Secret Intelligence Service (MI6) to become a double agent. After developing diabetes, Skripal was reassigned back to GRU headquarters in Moscow, where he passed on the identities of three hundred Russian agents to British Intelligence. Unfortunately for him, Russian spies inside MI6 had alerted GRU higher-ups to his spying. In December 2004 Skripal was arrested outside his house, and in a closed military court, convicted of high treason in the form of espionage. Stripped of his military rank and decorations, Skripal was sentenced to thirteen years in a high-security detention facility.

The Russians had Skripal, a double agent working for the British government, in their prison system. At the same time the American government had identified several Russian sleeper agents who were now imprisoned as spies within American maximum-security prisons. Naturally the Russians wanted their spies returned, and the British wanted Skripal returned to them. Diplomatic machinations between the British, Russian, and American governments resulted in an event that would not have been out of place in a John Le Carré Cold War spy novel.

On July 9, 2010, ten Russian sleeper agents on an American plane landed at Vienna's International Airport. As the spies disembarked they were welcomed back into Russian hands. In a tightly coordinated fashion, at the same time the American jet landed in Vienna, a plane from Russia carrying Skripal touched down at the Royal Air Force base in Brize Norton, England.[6] Skripal was now safely back with his British handlers. After an intense debriefing by MI6, Skripal eventually settled down in the city of Salisbury in southern England, where he hoped to live out his days with spy life far behind him. However, his past would eventually catch up with him.

In the early afternoon of March 4, 2018, Sergei and his thirty-three-year-old daughter, Yulia, walked out Sergei's front door, closing it behind them. They strolled toward the Mill pub for a drink, before heading on to a late lunch at an Italian restaurant. Shortly after leaving the restaurant, Skripal and his daughter started to feel queasy and experience blurred vision, as if something they had eaten had disagreed with them. Deciding to rest briefly before continuing to Sergei's home, the pair sat down at a nearby shopping center and waited for their nausea to subside.

By 4:15 p.m., the police received a phone call regarding two unconscious bodies slumped on a bench. Eyewitnesses reported that Yulia's eyes were wide open and staring blankly, with froth dribbling from her mouth. Sergei's whole body was stiff, his chin and clothes covered in the remnants of his vomit. Neither Sergei nor Yulia appeared to have any visible injuries, but they were clearly in a bad way. The ambulance service was called and father and daughter were taken to Salisbury District Hospital, where they remained unconscious and in critical condition.

At first medical staff suspected the Skripals were suffering from an opioid overdose, and started treating it as such, but with little effect. The first clue that the Skripals were suffering from something much more sinister occurred when Sergeant Bailey, a police officer who was the first to encounter the collapsed Skripals, was also admitted to the emergency room with similar but much less severe symptoms, including itchy eyes, rashes, and wheezing. Concern arose that the two unconscious individuals in intensive care were just the first two patients in what might become a disease epidemic. A breakthrough occurred when police learned of Sergei Skripal's former life as a Russian spy and double agent.

The medical team now realized that the Skripals were exhibiting the classic signs of organophosphorus poisoning—the toxic substance used in nerve agents. The Skripals were immediately treated with atropine, which bound and occluded the postsynapse acetylcholine receptors, preventing the excess acetylcholine from deadly overstimulation. Both patients were in a coma, with respiratory support to keep them breathing and prevent potential brain damage. All that could be done was to wait for their bodies to degrade and eliminate the nerve agent.

Now convinced that the Skripals had been intentionally poisoned, the doctors contacted experts at nearby Porton Down, the British government laboratory responsible for researching chemical weapons, including their detection and treatment. Examining samples from the Skripals, it was determined that they had been exposed to a chemical called Novichok (Новичóк Russian for "newcomer" or "beginner"), part of a series of nerve agents developed by the Soviet Union in the 1970s and '80s. The use of Novichok immediately threw suspicion on the Russian government, though Vladimir Putin, president of the Russian Federation, denied any involvement in the poisoning. Putin even argued that "if Russia had attempted to assassinate the double agent and his daughter, they would now be dead!" Such bluster impressed no one.

Yulia Skripal, likely because she was younger or had been exposed to a smaller amount of nerve agent, recovered faster than her father, and was released from the hospital under police protection. In a later interview Yulia remarked that she was stunned that "after 20 days in a coma, I woke to the news that we had been poisoned." Sergei remained in critical condition and unconscious for another month. Even when he regained consciousness, Sergei remained in the hospital for an additional three months before he, too, went into police protection in an undisclosed location.

Although all those subjected to Novichok were now out of hospital, questions remained as to how they had been exposed in the first place.

Eventually the culprits were discovered. On March 2, 2018, Alexander Petrov and Ruslan Boshirov had arrived at London's Gatwick Airport from Moscow, using passports supplied by the GRU. Petrov and Boshirov, both decorated colonels in the GRU, took up residence in the City Stay Hotel in the East End of London. Two days later they had traveled by train to Salisbury, where they were caught on closed-circuit TV footage near Skripal's house spraying liquid nerve agent on Sergei's front door. When Sergei and his daughter left their house and closed the door, they came into contact with the poison. Their job done, Petrov and Boshirov took the train back to London, and from there to Heathrow for a flight back to Moscow. Arrest warrants were issued for Alexander Petrov and Ruslan Boshirov, although these later turned out to be aliases, and the pair were actually Dr. Alexander Mishkin and Colonel Anatoliy Chepiga, both members of Skripal's old military intelligence unit, the GRU. Although the British government was certain it had enough evidence to charge the two men with conspiracy to commit murder, the two Russians maintained that they were simply tourists who got caught up in something beyond their control. An extradition request for the pair was considered, but officials argued whether there was any point, since Putin would likely refuse the request. Indeed during a press conference, Putin stated that not only were Petrov and

Boshirov innocent, but that they had identified the real culprits and he was waiting for them to turn themselves in.

The final victims of the Novichok narrative were Charlie Rowley and Dawn Sturgess. Charlie was excited to find a box of expensive perfume left in a charity bin near the center of Salisbury, a gift he was sure would impress his partner, Dawn. The box apparently contained a bottle of Premier Jour by Nina Ricci—but what was actually held within the small bottle was Novichok. The bottle had been used to transport and eventually spray Novichok onto the Skripals' front door. Instead, Dawn used it to spray the deadly chemical directly onto her wrists, likely exposing herself to ten times the dose received by the Skripals. Dawn died eight days later, the collateral victim of a bizarre assassination attempt.

ANCIENT POISON TO MODERN ANTIDOTE

The Romans were masters of personal and political murders using poison. So rampant were murders with drugs, such as belladonna, in first-century Rome, that laws were passed to suppress domestic poisonings. Juvenal, the Roman satirist best known for his cynical comment that "all we need from those who rule us are bread and circuses," also spoke of the deadly effect of belladonna, and said that the juice from a berry was much favored by wives wishing to rid themselves of unwanted husbands. By the time Paul Agutter decided to murder his wife, there was more than two thousand years' experience attesting to the utility of atropine as a murder weapon. Agutter took this as a sign that he, too, could use atropine to solve his problems.

Yet despite this lurid history, atropine finds uses in modern medicine. Treating spies poisoned with nerve agents is one of the more spectacular uses, but atropine is also used in hospitals to control heart rates, particularly in patients with a low heart rate, or even a stopped heart. Atropine can also be used to reduce saliva and airway secretions before surgery to prevent fluid getting into the lungs and causing pneumonia. What was once solely a poison has been rehabilitated as a therapy. In the next chapter we look at a drug that remarkably started out as a tonic but ended up being one of the world's nastiest poisons.

{ 3 }

Strychnine and the Lambeth Poisoner

Strychnine is a grand tonic, Kemp, to take the flabbiness out of a man.
—H. G. Wells, *The Invisible Man*, 1897

THE INVISIBLE MAN, *PSYCHO*, AND SHERLOCK HOLMES

It would seem bizarre that a drug such as strychnine, so entrenched in the mind as a deadly poison, would ever be seen as a tonic and pick-me-up, yet that is precisely how it was viewed until the beginning of the twentieth century. The title character, Dr. Griffin, in H. G. Wells's novel *The Invisible Man*, "found strychnine to be immensely beneficial," as Wells writes: "Griffin had a little breakdown. He started to have nightmares and was no longer interested in his work. But he took some strychnine and felt energized."

The benefits of strychnine seemed endless: Psychologist Karl Lashley found that strychnine enhanced the ability of rats to learn their way around mazes; it helped marathoner Thomas Hicks achieve Olympic Gold in 1904; medical students used strychnine as a pick-me-up while studying for examinations;[1] and even Adolf Hitler was reputed to have taken a strychnine tonic after the loss of German soldiers during the Battle of Stalingrad.

However, the use of strychnine also had a dark side, as reflected in its increasingly sinister depiction in popular culture. In Arthur Conan Doyle's novel *The Sign of Four*, Sherlock Holmes's stalwart companion Dr. Watson deduces murder by strychnine from the unusual grimace found on the victim's face. On the silver screen, Alfred Hitchcock would also use strychnine as a poison in *Psycho*, where Norman Bates used it to kill his mother, before rampaging through shower curtains with a kitchen knife. More recently, horror writer Stephen King invoked strychnine in his novel *Mr. Mercedes*, published in 2014.

Agatha Christie introduced strychnine as her favorite poison in her debut detective novel, *The Mysterious Affair at Styles*. So accurate was Christie's description of the effects of strychnine poisoning that a review of the novel in *The Pharmaceutical Journal* opined: "This novel has the rare merit of being correctly written—so well done, in fact, we are tempted to believe that the author had pharmaceutical training."[2]

Why have so many writers chosen to incorporate strychnine into their works of fiction? Likely because poisonings with strychnine have been so numerous and widely documented. Indeed, strychnine is listed third in the top ten poisons by number of criminal cases, behind only arsenic and cyanide.

THE STORY OF STRYCHNINE

Strychnine, like atropine (see chapter 2) and as well as caffeine, nicotine, and even cocaine, is a plant alkaloid. All these compounds are bitter-tasting chemicals that plants generally put into parts that they don't want to be eaten. Ironically, humans have gone to great lengths to harvest plants to get as much of those same alkaloids as possible. Strychnine comes from plants of the genus *Strychnos*; it is a term coined by master labeler Carl Linnaeus in 1753. Although strychnine is found in all *Strychnos* species, it is most abundant in *Strychnos nux vomica*, which sounds slightly more scientific in Latin than its common name, the Asian vomit button tree. The strychnine tree is an evergreen, native to India, Sri Lanka, Tibet, southern China, and Vietnam. Although the tree grows quite commonly in Asia, there are very few stories of it being used for nefarious purposes within these regions. Whether this reflects a true reluctance of the local populace to use strychnine to kill, or simply a lack of record keeping, is not clear, but strychnine has been used in Asia mainly for controlling pests such as rats.

Strychnine entered the European market when ships began trading with the wider world. All ships have rats, and sailors do not appreciate rats eating their food or spreading disease. Thus strychnine was a popular rodent control solution among merchant vessels. By the late 1800s nearly five hundred tons of *Strychnos* seeds per year were being imported into London, most of it used to poison vermin such as rats and mice. Although it was difficult to purchase strychnine from a pharmacist, it could be easily obtained by the general public in threepenny and sixpenny packets, under the name Vermin Killer.

Butler's Vermin Killer consisted of a mixture of flour, soot, and strychnine, to be applied to a piece of bread or cheese and left on the kitchen floor overnight. So rapid and effective was strychnine as a rodent killer that mice and rats would often be found dead close by the poison. In Alfred Swaine Taylor's *Manual of Medical Jurisprudence*, published in 1897, Taylor remarked of vermin killers that "the powders are a fertile source of poisoning, either through accident or design; they are openly sold by ignorant people to others still more ignorant, and are much used for suicidal purposes." So readily available was strychnine that, marketed as "Vermin Powders," its purchase didn't warrant any raised eyebrows, even if the buyer had a murderous intent.

The use of Vermin Powders was not confined to the elimination of rodents, but was also liberally used to get rid of stray dogs and cats. One such purchaser was writer Henry F. Randolph, who in May 1892, bought some strychnine to poison an annoying cat. As any sensible nineteenth-century writer would do, he elected to put the poison in a drawer by his bedside rather than outside, in—say—a garden shed, in a container marked "POISON." One night Randolph awoke and decided to take a dose of quinine, another bitter alkaloid. Not surprisingly, in the dark, he picked up the bottle of strychnine and took some of that instead; three and a half hours later he was dead. It is safe to assume that the moral of this story is that keeping poison on your bedside table is not the most sensible idea.

Like all plant alkaloids, strychnine has a bitter taste. In fact, one of strychnine's claims to fame is that it is the bitterest substance known to man, and all other bitter tastes are ranked relative to it.[3] Another plant alkaloid known for its bitter taste is quinine, and as seen in the previous chapter, quinine has been used as a treatment for malaria, as well as a therapy for leg cramps. French physicians during the Napoleonic era used somewhat dubious logic to assume that if quinine, which is a bitter white powder, was medically useful, then any other powder that was also white and tasted bitter must necessarily also be beneficial. This logic ensured that patients with various ailments, including malaria, could be treated with strychnine just as easily as with quinine. Fortunately for scores of French patients, it was soon realized that this approach to drug discovery was not quite as useful as first thought, and the practice of treating all diseases with random bitter white powders came to an end.

THE LAMBETH POISONER

By 1891 Victorian London was recovering from Jack the Ripper's reign of terror. It had been three years since the last Ripper murder in 1889, and life for London's prostitutes was returning to some form of normality. That is, until the dead body of a working girl, nineteen-year-old Ellen Donworth, was found on the streets of Lambeth. Had Jack returned? Or was a new killer preying on the ladies of the night?

Life for Lambeth's prostitutes was tough. Even though they could earn ten or twelve times what they would make as domestic servants or working in a factory, their lives were nasty, cruel, and short. There were very few brothels in Lambeth, and so the women carried out their trade either in their own homes or out on the streets, making them susceptible to violent clients. Ellen Donworth had given up her job labeling bottles to become a prostitute. On October 13, 1881, Ellen received a note from a man calling himself Fred, asking her to meet him at the nearby York Hotel. Fred turned out to be a very attractive gentleman, wearing a silk-lined cape and a silk top hat, and carrying a gold-topped cane. Ellen was very impressed, sensing that the gentleman was wealthy, and hoped he might become a regular client. At around seven that evening she said good-bye to Fred and left the hotel.

Within minutes Ellen was finding it difficult to walk and suffering from agonizing stomach pains. She was found by a friend who thought she was drunk—but Ellen was affected by something much more sinister than alcohol. Taken home and placed in her bed, Ellen was gripped by terrifying convulsions, every muscle in her body contracting simultaneously, causing her back to arch horrifically. Between screams of agony, Ellen told her landlady that she had drunk twice from a bottle of pale liquid given to her by "a tall gentleman with cross eyes, a silk hat and bushy whiskers" who went by the name Fred. Racked with pain, Ellen was taken by cab to the hospital, but she died on the way. A postmortem revealed that her stomach contained large quantities of strychnine.

A popular venue for entertainment in mid-nineteenth-century Britain was the music hall. Here, the public was treated to a variety of performances from theatrics to acrobatics, comedy and popular (sometimes bawdy) songs of the day. Among the grander of the music halls were the Alhambra and the St. James. One of the more seedy aspects of some Victorian music halls was their use by prostitutes to pick up clients. A week after Ellen Donworth's death, Fred was seen at the Alhambra with another prostitute, Louisa Harvey. After an evening at the variety show, Fred had taken Louisa to a hotel in Soho, where they spent the night. In the morning he made an appointment with her for eight o'clock that evening, so that he could give her some pills to take care of acne on her forehead. At the prearranged time, Fred and Louisa met across from the entrance to the Charing Cross Underground station. After a few drinks at a nearby pub, the couple walked down to the Embankment, a pedestrian walkway running alongside the River Thames. As they strolled along, Fred removed two white pills, wrapped in tissue paper, from his waistcoat pocket and pestered Louisa to swallow them. As soon as Fred was convinced that Louisa had consumed the pills he turned on his heels and walked off into the London night.

For several months Fred seemed to have vanished, but he would return to claim two more victims. On April 11, 1892, two women living in Stamford Street, twenty-one-year-old Alice Marsh and eighteen-year-old Emma Shrivel, were having a dinner of canned salmon at their home with a man they had recently met. Later one of the girls was found foaming at the mouth, her roommate lying groaning on the bed. When they were asked what had happened, they replied that they had swallowed pills given to them by their visitor. Asked why they would take pills from a stranger, they revealed he was no stranger: He was a doctor. Both women writhed in excruciating agony for several hours before they died.

Three prostitutes had been horribly poisoned, and a fourth had disappeared without a trace. The killer was still on the loose, and although his identity was still unknown, the press had already given him a name: "The Lambeth Poisoner."

Though the London press had yet to discover it, the killer was already known as a poisoner, and an abortionist, on the other side of the Atlantic Ocean. Thomas Neill Cream, a Scottish immigrant, graduated with an honors degree in Medicine from McGill University in Montreal in 1876. With above-average intelligence, good looks, and charm, Cream was always very popular with the womenfolk. While a student at McGill, he had a torrid affair with Flora Brooks, which resulted in Flora becoming pregnant. Rather than go through with the pregnancy out of wedlock, Cream decided that he would abort the baby himself, almost killing Flora in the process. Flora's father, Lyman Henry Brooks, a wealthy hotelier, was not impressed by Cream's treatment of his daughter, and persuaded him—at gunpoint—to save his daughter's honor by marrying her. Cream dutifully wed Flora on September 11, 1876, but had no intention of really being married to her; Cream promptly left the country the next day, sailing to England to begin further postgraduate medical training at St. Thomas's Hospital School in London.

Cream's time in London was not too successful, as he failed

to obtain his MRCS (Member of the Royal College of Surgeons) certification, and so he transferred to the Royal College of Physicians and Surgeons in Edinburgh, Scotland, where he finally managed to qualify to practice. One of Cream's classmates was a young medical student called Arthur. While Cream would attain notoriety as a serial murderer, Arthur, whose full name was Arthur Conan Doyle, would achieve fame capturing criminals in the form of his creation, Sherlock Holmes.

While studying in Edinburgh, Cream received news that his wife was ill, and generously sent her some medication. Mrs. Cream died soon after, and her death was reportedly caused by consumption (tuberculosis). Subsequent events would suggest that she had died from consuming her husband's poison.

It wasn't long after Cream received his newly minted Edinburgh medical license before one of his patients died. In 1879 a young pregnant woman was seen walking into Cream's office. What happened next will never be known, but the next time his patient was seen was when she was found dead in a shed behind the office, poisoned by a chloroform overdose. Although Cream avoided murder charges, suspicions of incompetence and malpractice ruined his reputation, and he decided that a trip back across the Atlantic was in his best interests.

In October 1871 Chicago had been devastated by a great fire that killed around three hundred people and destroyed more than three square miles of buildings. Eight years later, when Cream arrived in Chicago, the city was beginning to recover and rebuild through an influx of immigrants. Cream set up practice near the city's red-light district, where he found he could make some fast money performing abortions for the local brothels. By 1880 Cream's activities were well known in the red-light district.

To be fair to Cream, most abortionists at the time carried out their operations in a manner more akin to butchery than to medicine. Many patients simply bled to death after botched operations, or contracted life-ending infections from dirty surgical instruments. Cream was known to be aided in his abortions by an African American midwife by the name of Hattie Mack. Her mysterious disappearance one day aroused the suspicions of her friends, who notified police of her absence. A search of Hattie Mack's rooms revealed the presence of the decomposing corpse of a young prostitute called Mary Ann Faulkner, who appeared to have bled to death. Despite Hattie Mack's attempted escape, the police finally caught up with her, but since it was Cream the police were really after, they persuaded Hattie Mack to give evidence against him in exchange for leniency. Determined to save her own neck, Mack was more than happy to provide the police with any information they wanted, telling them that Dr. Cream had performed as many as fifteen abortions from a single brothel, and had boasted of carrying out at least five hundred abortions in total.

This time Cream was arrested. But although the police tried to make a solid case against him, Cream convinced the coroner that it was in fact Hattie Mack who was responsible for Faulkner's death. After all, it was in Mack's apartment that Faulkner's corpse was discovered. Moreover, how could a jury possibly take the word of a female midwife, and known abortion provider, over that of the charming and well-mannered Dr. Cream? Cream was found not guilty of murder.

Ironically, of all the women that Cream could have been found guilty of murdering, it was the murder of a man that finally sent him to prison.

In 1881 sixty-one-year-old Daniel Stott was a passenger agent for the Chicago and North Western Railroad in Garden Prairie, a town seventy miles northwest of Chicago. Stott lived with his young wife, thirty-three-year-old Julia, and ten-year-old daughter, Revell. For the most part Stott lived a comfortable life, but he was prone to attacks of epilepsy, and was in declining health. But suddenly news came from Chicago that a wonderful "foolproof" remedy for epilepsy was now available from a Dr. Thomas Neill Cream. Stott made the train journey to Chicago, where he visited Cream and was prescribed his new patent medicine. Whether the prescription provided any relief is doubtful, but Stott was convinced enough to send his wife back to Chicago whenever his medication ran out. Cream quickly charmed and seduced Julia, and the two started a torrid affair.

On the Saturday morning of June 11, 1881, Stott kissed his wife good-bye as she boarded the train to Chicago to visit Cream. The next day Julia returned with the precious medicine and prepared a dose for her husband. John Edgecomb, a blacksmith and close friend of Stott, called in to see his friend just as Julia was preparing her husband's medication. Edgecomb would later testify that he "watched Daniel Stott go into a state of convulsion and die, while his wife remained unmoved and unwilling to summon a doctor." Despite this, Stott's death occasioned little surprise, as he had been ill for some time. The funeral went as planned, and the widow, dressed in black, wept appropriately at the cemetery.

Cream appeared to have gotten away with murder again; but, in a startling display of hubris, he sent a telegram to the district attorney hinting that Stott had not died from natural causes, but rather from strychnine poisoning. Cream followed up the telegram with letters to the coroner demanding an exhumation of Stott's body and a postmortem examination. When a test of Stott's medicine on a stray dog proved fatal, authorities launched a full-scale investigation. The coroner's jury found that Daniel Stott had indeed been murdered, implicating not only his widow but also Dr. Cream.

At the trial Julia Stott proclaimed her innocence, and that she had no idea the medicine was poison. She further testified that "Dr. Cream told me that he was working on a plan to poison people so that he could sue some Chicago drug stores, but I didn't think he would poison Dan [my husband]."

Cream nonchalantly took to the stand in his own defense, denying all charges and maintaining his complete innocence: "I firmly believe that Mrs. Stott herself killed her husband," Cream declared. "Mr. Stott came to me some time before his death and told me that his wife had been criminally with another man. He asked my help. The next time Mrs. Stott came to Chicago I told her what her husband had said. She flared up and screamed: Damn him, I'll give a dose that will fix him!" During closing arguments, Cream read a newspaper, looking up at the jury with an occasional smile.

The jury took only three hours' deliberation to find Cream guilty of murder. Julia Stott walked free after agreeing to testify for the prosecution. The judge sentenced Dr. Cream to Joliet Prison in Illinois "for the rest of his natural life." Several months after the trial, a new tombstone appeared on Stott's grave. Stott had been a member of the Belvidere Masonic Lodge, and rumors quickly surfaced that lodge members had erected the stone under cover of darkness. Inscribed on the tombstone were the words: "Daniel Stott died June 12, 1881, age 61 years: poisoned by his wife and Dr. Cream."

Although Cream was sentenced to life in prison, he was not to languish in an American jail forever. In 1891, after only ten years, Cream's sentence was reduced to seventeen years by Joseph W. Fifer, governor of Illinois, which, with time off for good behavior, meant that Cream could be released almost immediately. It would later be argued that money had changed hands to bring about the governor's reduction in Cream's sentence and subsequent release. Cream had inherited sixteen thousand dollars on the death of his father, roughly equivalent to some four hundred thousand today. Given the penchant for political corruption in Illinois and Chicago (four Illinois governors have been sentenced to jail), this story likely has some truth to it.

Upon his release Cream decided to travel back to England, arriving at the port of Liverpool on October 1, 1891. A train ride took him to London, where he set up residence at 130 Lambeth Palace Road, in the Waterloo area of London. Though he was free again, a decade in prison had taken its toll on Cream, and two things took center stage in his mind: feeding his prison-acquired drug habit, and solving the problem of what he referred to as "streetwalkers."

Soon after Cream's arrival in London, three prostitutes had died in a horrible way from strychnine poisoning, and one had disappeared without trace; the poisoner was still at large.

The hunt for the "Lambeth Poisoner" was now front-page headlines, and the public couldn't get enough of the story. People all over London were trying to solve the case themselves. John Haynes was a former New York detective who was now living in London and desperate to get a job at Scotland Yard by cracking the Lambeth Poisoner case. Haynes became friendly with Dr. Thomas Neill, a physician interested in the poisoning aspects of the murders. Many hours were spent at pubs discussing the available evidence, and suggesting likely suspects—and it was these conversations that would eventually unlock the case.

Dr. Neill started discussing things about the case that were unknown to Haynes, and unknown even to Scotland Yard. Dr. Neill even took Haynes on a tour of some of the murder sites. Remarkably, Dr. Neill informed Haynes that he had actually known three of the poisoned prostitutes: Louisa Harvey, Matilda Clover, and Ellen Donworth. This seemed strange, since at that time, it was thought that Matilda Clover had died from alcoholism and not poisoning, and Louisa Harvey had just disappeared; there was no evidence she was dead. Haynes's suspicions about Dr. Neill, and his eagerness for career advancement, led him to Scotland Yard, where the detectives were very interested in what he had to say.

House-to-house inquiries in Matilda Clover's neighborhood

uncovered a witness to her last moments, who described Matilda's agonizing spasms and arched back, strikingly similar to the symptoms experienced by Ellen Donworth, Alice Marsh, and Emma Shrivel. Further interviews with Clover's friends revealed that they had seen her in the company of Dr. Neill shortly before her death. The mounting evidence led to the exhumation of Matilda Clover's body on May 6, 1892. The pathologist in charge of Clover's postmortem was very thorough, taking more than three weeks for his investigation. Excising pieces of Clover's liver and stomach, he ground them to a paste, forming a liquid he could test. Tasting the liquid, the pathologist noted its bitterness, a clear indication of an alkaloid—but which one? The pathologist administered some of the liquid extract to a frog, which shortly displayed the classic strychnine convulsions before it died. Matilda Clover was clearly another victim of the Lambeth Poisoner, a fact already disclosed by Dr. Neill. But how was Dr. Neill already privy to this knowledge?

This was a question Scotland Yard was eager to have answered. Matilda Clover's inquest took place on June 22, with the jury not only determining that Matilda Clover had died from strychnine poisoning but also naming Dr. Neill as the person unlawfully causing that death. While Dr. Neill was in police custody, the authorities received interesting information from Canada and Illinois regarding the true identity of the man they were questioning. Dr. Thomas Neill was in fact Dr. Thomas Neill Cream. With this information in hand, Cream's real character as an abortionist and convicted murderer was revealed.

Dr. Cream was now on trial for the murder of Matilda Clover.

Under the British legal system, only evidence directly relevant to her murder could be presented. However, the presiding judge at the Old Bailey Court, Sir Henry Hawkins, gave the prosecution wide scope in allowing evidence of Cream's past crimes to be brought into evidence. Even though Cream was on trial for only Matilda Clover's murder, the prosecution argued that the murders of Ellen Donworth, Alice Marsh, Emma Shrivel, and Louisa Harvey showed that Cream was systematically poisoning prostitutes with strychnine.

What happened next was one of the most extraordinary events in British legal history: A surprise witness was brought in to appear for the prosecution. The bombshell witness was none other than presumed murder victim Louisa Harvey. Harvey had read her name in the newspaper in connection with the trial, and was now on the witness stand to give a clear account of Cream's attempt to murder her. Harvey described how Cream had given her pills to take, and she had pretended to swallow them, throwing them on the ground at the last minute. Both the jury and Cream were completely stunned. There before Cream was the woman he assumed he had murdered months before.

The jury took just ten minutes to find that Matilda Clover had died of strychnine poisoning, and that the strychnine had been administered by Cream with the intent to murder. Cream was also found guilty of the murders of Donworth, Marsh, and Shrivel, as well as the attempted murder of Louisa Harvey. Less than eighteen months after his release from the Illinois State Penitentiary, Thomas Neill Cream was hanged at Newgate Prison on November 15, 1892.[4] A report in the *Canadian Medical Associa-*

tion Journal describing Cream said "he was a drug fiend, and this may have been a factor in his career of habitual murder. He used his medical knowledge to slay his unfortunate victims. What suffering would have been avoided had not soft-hearted, misguided enthusiasts sought his release from Joliet Prison."

HOW STRYCHNINE KILLS

Of all the poisons used to commit murder, strychnine is probably one of the most insidious. It leads to an agonizing death for the victim, and a frightful sight for witnesses, who are unable to provide any aid or comfort. Strychnine tortures its victims by racking their body with excruciating spasms, before allowing death finally to rescue them from an earthly hell. Within a few minutes of exposure to strychnine, either by injection, inhalation, or ingestion, the first symptoms start to appear, with generalized twitching of the muscles and tightness in the limbs. The muscles of the jaw spasm and tighten, leading to lockjaw, and the other facial muscles force the mouth into a grotesque exaggerated grin, known as *risus sardonicus*, or "sardonic grin." Within minutes the muscles of the rest of the body start to spasm and contract uncontrollably, occurring in waves lasting three to four minutes before there is some relief, only to be assaulted by another wave of contractions a few minutes later. Death finally arrives a few hours after exposure to strychnine.

In humans, the muscles of the back are usually much stronger than those of the abdomen, and so sustained contractions of both the back and abdominal muscles following strychnine poisoning often lead to a rigid, arched back. This causes the victim's whole body to rest only on the back of the head and the heels, in what is referred to as *opisthotonos*.

At Cream's trial one of Matilda Clover's housemates recalled the events surrounding Clover's death. "After I had gone to sleep I was woke up by hearing screams. I slept in the back room under that where Clover slept." The witness said that she woke up the landlady and the pair went to Clover's room. "She was lying across the bed with her head between the mattress and the wall. She was screaming in great pain. There were times when she seemed to have relief, and then the fit came on again. Her body was all of a twitch."[5] To the police and prosecution, such terrifying symptoms observed by the unfortunate witness to Matilda Clover's demise clearly connected her death to the rampage of Dr. Cream's poisonings.

As mentioned earlier in this chapter, strychnine was a favorite of Agatha Christie's for the dramatic characteristic of these spasms. In her first detective novel, *The Mysterious Affair at Styles*, she described the striking demise of Mrs. Emily Inglethorp, in which "a final convulsion lifted her from the bed, until she appeared to rest upon her head and her heels, and her body arched in an extraordinary manner."

Often strychnine victims have a very ruddy complexion, as blood vessels open in response to oxygen deprivation in the overworked muscles. As the poisoning progresses, the heartbeat becomes erratic, causing elevated blood pressure and rapid breathing. Death ultimately results from asphyxiation, as the muscles of the diaphragm become exhausted and stop contracting. Fully conscious, the victim tries to will his or her body to breathe in life-sustaining oxygen, but the muscles, totally worn out, are unable to comply. One of the cruelest aspects of strychnine poisoning is that victims have heightened senses, making them even more profoundly aware of their downward spiral to death.

Strychnine affects the nerves of the central nervous system (CNS), the nerve network that sends messages from the brain to the body, and receives signals from around the body that are sent back to the brain. In the CNS, one particular set of nerves are called "motor neurons," and as the name suggests are involved in sending signals to muscles telling them to do work, to turn the pages of this book, for example, or get out of a chair to make a cup of tea. The intensity of signals down motor neurons is not steady, but can be increased or decreased to vary the strength of the muscle contraction. Think of music played on a radio or cell phone; the volume can be made loud or quiet by the turn of a dial. The same is true in the CNS, where neurochemicals either help amplify the signal or diminish it. This allows our same hand and arm muscles to gently hold a baby or strongly grasp a sledgehammer.

One of the ways in which signal intensity is minimized is through a small chemical called glycine. Glycine is the smallest of all the amino acids, and can be thought of as a brake, diminishing the intensity of a signal down the motor neuron. Lodged in the membranes of nerves are special proteins called glycine receptors, which recognize and hold on tightly to the glycine molecule. When glycine binds to nerves, it makes it much harder for the nerve to send a strong signal down its length, making the signal and the subsequent muscle contraction weaker. This is a good thing, as it means that the nerves don't just send signals at the slightest provocation, but require a clear message to cause a muscle to contract.

Strychnine attaches to the glycine receptor three times more strongly than glycine, but whereas glycine causes a moderation of the signal, strychnine triggers an amplification of the message, sending strong instructions to muscles for prolonged intense contraction at the slightest signal. If the muscles happen to be jaw muscles, the jaw will clench and produce lockjaw. The merest brain activity will cause the muscles of the back and abdomen to contract, giving rise to the classic *opisthotonos* symptom of strychnine poisoning. Muscle spasms typically occur in waves, each becoming more violent than the last, until the diaphragm muscles wear out, breathing stops, and the victim dies. As if the uncontrolled muscle spasms are not bad enough, the enhanced, amplified signals coming from the ears and eyes make the victims even more aware of their surroundings and what is happening to their bodies. The effects of strychnine poisoning are so rapid and dramatic that, unless treatment is administered immediately following exposure, the chances of survival are negligible.

If the effects of strychnine poisoning are so terrible, why was it ever considered a tonic? A drug that causes muscles to convulse to the point of ripping tendons apart would seem an unlikely performance-enhancing drug. But as we saw in chapter 1, Paracelsus's statement that the dose makes the poison suggests that low doses of strychnine may enhance athletic performance by enhancing muscle contractions. Low doses of strychnine certainly do increase swimming in lampreys and tadpoles, but whether this extends to increased performance in humans is less than certain. When Thomas Hicks collapsed after crossing the finish line of the 1904 Olympic marathon, it was not clear whether his success was due to the 2 mg of strychnine his manager had given him or the plentiful supply of brandy he consumed during the race.

Remarkably, Hicks was not the last Olympian to use strychnine as a performance enhancer. Chinese volleyball player Wu Dan was disqualified after testing positive for strychnine in the 1992 Barcelona Olympics, and Kyrgyzstan weight lifter Izzat Artykov was stripped of his Olympic bronze medal at the 2016 Rio Olympics, when strychnine was found in his system. The likelihood that strychnine could ever be a real tonic is probably more snake oil than science.

A CURE FOR STRYCHNINE

Unfortunately there is no specific antidote to strychnine poisoning, and treatments instead focus on alleviating symptoms. Since the motor neurons have become hypersensitive, keeping the patient calm and in a dimly lit room can sometimes help prevent the nerves from firing inappropriately. Muscle relaxants can also stop the convulsions, allowing the body to rid itself of the poison slowly. Since these relaxants would also relax the diaphragm, however, the patient must be put on artificial respiration for several hours. Diazepam, better known as Valium, can be used to calm the nerves of the central nervous system, and to relax the muscles, suppressing the worst of the convulsions. Although not a specific treatment for strychnine poisoning, activated charcoal can be given to patients to help soak up any remaining strychnine in the stomach and intestines, and prevent it being absorbed by the body. Activated charcoal is essentially a large lump of charcoal full of holes, which strychnine, or indeed almost any other toxic drug or chemical, can enter and be trapped.

Today clinical trials evaluating new drugs for use in patients are subject to very stringent regulations. All sorts of health guards must be in place to protect the volunteers who will be testing the safety and efficacy of new therapies. Things were a little more lax in nineteenth-century France, however, when Professor Pierre Tourey, a pharmacist from Montpellier, tested the effects of activated charcoal on strychnine poisoning. In 1831 Tourey performed a demonstration in front of the French Academy of Medicine, consuming ten times the lethal dose of strychnine, mixed with fifteen grams of charcoal. Whether certain of his own convictions or simply foolhardy, Tourey survived his self-imposed strychnine poisoning. One might imagine he would be hailed a hero for demonstrating a treatment for strychnine; in fact Tourey was booed off the stage by unconvinced academicians.

Tourey no doubt based his test on the work of another Frenchman, Michel Bertrand, who had appeared before the same French Academy of Medicine some eighteen years before. In that case Bertrand was extolling the virtues of charcoal for arsenic poisoning. He swallowed five grams of arsenic trioxide (forty times the lethal dose) along with some charcoal. He survived the episode unscathed, exhibiting none of the classic symptoms of arsenic poisoning. Exactly what nineteenth-century French doctors had against charcoal is not clear, but those early demonstrations did provide proof that charcoal was useful in treating poisons, and even today is used for mopping up poisons, and overdoses of pills, that have been swallowed.

In this chapter we saw how a supposed tonic was used as a toxin by Dr. Cream, landing him in a murder trial at the Old Bailey (London's Central Criminal Court). The next chapter will see two more trials at the Old Bailey. Both murderers used the same poison to kill their victims, and both trials generated an enthralled nationwide interest that was stoked by the press. One big difference between the trials: They took place 130 years apart.

. { 4 } .

Aconite and Mrs. Singh's Curry

What is the difference, Potter, between monkshood and wolfsbane?
—J. K. Rowling, *Harry Potter and the Sorcerer's Stone*, 1997

A SHORT HISTORY OF ACONITE

The *Better Homes and Gardens* website asks, "How can you not fall in love with a perennial that has regal blue spires?" Indeed, monkshood is quite an attractive plant, with tall spikes of hooded purple or blue blooms that flower in late summer to fall. Its name comes from the observation that the flowers look rather like the cowls worn by medieval monks. However, monkshood is not the only name by which the plant is known; throughout history it has garnered rather more sinister appellations, including wolfsbane, leopards bane, and devil's helmet. The word *bane* means "poison," and refers to the use of the plant as an arrow poison for hunting wolves and other dangerous carnivores. Not only is monkshood, or wolfsbane, dangerous to wolves, it is also quite deadly for humans, appropriately earning the moniker "Queen of Poisons."

The word *aconitum* may come from the Greek word *Ακόντιο*, meaning "sharp dart" or "javelin," the tips of which were coated with poison; or from *akonae*, because of the rocky ground on which the plant was thought to grow. In the *Iliad*, written in 762 BC, Homer describes Hercules' test to capture Cerberus, the monstrous three-headed dog, from the underworld, and bring it up into the world of the living. As Hercules subdued the horrific beast, the noxious drool from its three snarling snouts fell upon the ground, and immediately poisonous aconite plants sprang up.

The genus *Aconitum*, or aconite, comprises more than two hundred species of flowering plants that grow in damp and partly shaded areas of Europe, Asia, and North America. All these plants contain the alkaloid aconitine, which, like other plant alkaloids, is not necessary for the plant's growth, but acts as a deterrent system to the plant being eaten. Most of the aconitine is found in the roots of the plants, but all parts of the plant can be deadly if eaten. Accidental ingestion of the roots actually turns out to be far more common than would be expected, since they are often mistaken for horseradish. In 1856 a dinner party was held in Dingwall, a Scottish village thirty miles north of the famed Loch Ness. A servant had been sent into the back garden to dig up some horseradish to make a sauce for the roast beef dinner. Instead, the unwitting servant dug up some aconite, and the cook, unaware of the difference, blithely grated the aconite root into the sauce. The poisoned dinner promptly killed two priests, who were guests at the party; other guests who ate less were sickened but survived.

An October 1882 issue of the *British Medical Journal* contained a bizarre article reporting on a man who saw something drop from a passing van and, thinking it was horseradish, not only ate some but gave a piece each to three other men and his sister. Within a very short time all five were in the hospital complaining of a numb sensation in the mouth, with partial paralysis of their arms and legs. Artificial respiration was maintained for four hours before the symptoms gradually wore off and the patients recovered. The supposed horseradish was found to be aconite root.

Aconite plants have been used for centuries in herbal medicines as a treatment for gout, probably due to the pain-relieving local anesthetic properties of its extracts. In the nineteenth century, ointments and liniments made from aconite were used by physicians for all kinds of ailments, including rheumatism, neuralgia, sciatica, migraines, and even toothaches. Powdered wolfsbane was in fact used by dentists to numb aching cavities in patients before the advent of novocaine or lidocaine. Fortunately a trip to the dentist these days does not rely on such ancient painkillers.

Although aconite alkaloids do have anesthetic properties, the margin of error between numbing a pain and killing the patient was very narrow. In 1880 one doctor prescribed aconite drops to a young boy. Shortly after taking the medicine the boy became extremely ill with chills and convulsions. The boy's mother promptly returned to the doctor, accusing him and his prescription of harming her son. Incensed that anyone, particularly a woman, would dare question his skills, he took a dose from the boy's medicine bottle to demonstrate that it was perfectly safe. Five hours later the good doctor was dead from aconite poisoning.

Where most doctors were content simply to prescribe aconite as a painkiller, one doctor and professor became increasingly concerned about the use of aconite as a poison. Though his interest was purely academic, one of his students would put the theories into practice.

THE PERFECT MURDER

Sir Robert Christison was a professor of Medicine at the University of Edinburgh for more than fifty years, eventually serving as president of the Royal College of Physicians at Edinburgh. During his time in Edinburgh he became fascinated by poisonings and toxicology, writing a hugely popular textbook called *A Treatise on Poisoning*, which ran to four editions. His interests also led him to medical jurisprudence, and he was often called upon as an expert witness for the prosecution in murder cases. During one particular trial, he was being cross-examined on the ease of detecting poisons in a dead body. Christison turned to the judge and said: "My Lord, there is but one deadly agent of this kind which we cannot satisfactorily trace in the human body after death; and that is—"

The judge at once interrupted him. "Stop! Stop! Mr. Christison, please," the judge cried. "It is much better that the public should not know it!"

In a subsequent lecture to Edinburgh medical students, Christison revealed that what he had been about to say before the judge interrupted him was that "the perfect undetectable poison" was aconite. Several of the renowned doctor's students would later report that there was one member of the class who was exceptionally diligent in taking notes when Christison was discussing aconite. We will meet this pupil shortly.

DR. LAMSON'S DUNDEE CAKE

"I looked up tontine in the dictionary," said Lucy.

"I thought you probably would," said Miss Marple equably.

Lucy spoke slowly, quoting the words. "'Lorenzo Tonti, Italian banker, originator, 1653, of a form of annuity in which the shares of subscribers who die are added to the profit shares of the survivors.' A will of that kind, ending so that if there was only one survivor left he'd get the lot . . ."

—Agatha Christie, *4:50 from Paddington,* 1957

In the nineteenth century, one man went on a killing spree right out of a Christie mystery as he worked his way through his in-laws' family to get his hands on their inheritance. In 1852, George Lamson was born in New York to clergyman William Lamson and his wife, Julia. While George was still young, the family sailed across the Atlantic to live in England. George was always bright, and at age eighteen he enrolled as a medical student at the prestigious University of Edinburgh. After graduating, the young Dr. Lamson served as an army surgeon during the various wars that ravaged Eastern Europe and the Balkans at the end of the nineteenth century. For eight years Lamson appeared to serve with distinction, earning the Legion of Honor for his work in the Franco-Prussian war, the first of many decorations. Upon returning to England, a chest full of medals was not the only thing Lamson brought back with him: he also returned with a secret opiate addiction.

In 1878, on the Isle of Wight, a small island off the southern coast of Britain much favored by Queen Victoria, Dr. Lamson married a Welsh girl by the name of Kate George John. Kate was the twenty-five-year-old orphaned daughter of a wealthy linen merchant, who, upon marriage, was entitled to claim her portion of her parents' legacy; a legacy that, under Victorian law, automatically also became her husband's. Kate was one of four siblings who shared equally in the inheritance from their parents. Kate's sister had married and acquired her share the previous year. There were also two brothers, Hubert and Percy, but as they were still minors, their shares of the inheritance were managed for them by a trust. In a classic tontine arrangement, if any sibling were to die before marriage or before reaching the age of majority at twenty-one, his or her share of the wealth would be equally distributed among the remaining siblings.

In 1880, using some of the money left from Kate's inheritance, Lamson purchased a medical practice in the seaside town of Bournemouth. Recent research conducted at Bournemouth University has revealed that the town was well favored by upper-class drug addicts as a quiet place to feed their addiction. Records from Parr's Pharmacy, just a few yards from the Lamsons' home, showed it regularly dispensed morphine to hotel visitors. At first, Lamson was able to conceal his addiction, becoming a pillar of the community and even obtaining a commission in the First Bournemouth and Hampshire Artillery Volunteers based on his distinguished military career. But despite his apparently successful medical practice, Lamson's drug addiction was eating into his assets, and he was running up huge debts as he strove to maintain his wealthy lifestyle. His landlord was left £40 pounds (almost £5,000 or nearly $7,000 today) out of pocket for unpaid rent, and he was just one of many. Struggling to feed his addiction, Lamson pawned his watch and medical instruments to gain cash. He tried to borrow money from acquaintances, and obtain cash advances on checks that would always bounce. Eventually the Bournemouth bank stopped honoring any of his checks, leaving a trail of money owed to porters, accountants, wine merchants, and even strangers. Desperate for money, his mind addled by the morphine addiction, Lamson turned his thoughts to his in-laws' inheritance. What Lamson desperately needed was for his brothers-in-law to die.

In June 1879 Lamson seemed to receive some cooperation in this area, as Hubert died suddenly and unexpectedly, leaving his share of the inheritance—some £3,000, about £370,000 (around $518,000) today—to be split equally between the remaining siblings. The surviving brother-in-law, Percy Malcolm John, was nineteen years old and suffered from severe curvature of the spine, forcing him to use a wheelchair or be carried. Although he had no use of his legs, his upper body was fine, and he was generally in good health. Money from his parents' will allowed Percy to attend school at Blenheim House in Wimbledon, London. Percy was getting precariously close to the age of twenty-one, in which case he could avoid the tontine altogether. However, if Percy were to die unexpectedly before his majority, his £3,000 ($4,000) would be equally distributed between his two sisters, meaning that Lamson would have immediate access to £1,500 (about £188,000 or $260,000 today), an amount that would go a long way toward solving his financial problems. This realization on Lamson's part sealed Percy's fate.

Lamson decided to eliminate the only obstacle remaining between him and his in-laws' money. The first step was to acquire the right poison, and Lamson bought 2 grains (approximately 130 mg) of aconite from a pharmacist in London for around £12.50 (£1,500 or $2,073 today). Fortunately for Lamson, when the pharmacist found out that he was a physician, he declined to ask any potentially awkward questions as to what Lamson intended to do with the drug, assuming Lamson was giving it to a patient for pain relief. Next Lamson wrote Percy a letter stating that he would shortly be going abroad, and would like to meet with Percy before he left.

On the evening of December 3, 1881, Lamson arrived at Percy's boarding school. As he was waiting in the dining room for Percy to be brought up the stairs, Lamson took out a Dundee cake, a traditional Scottish fruitcake he had brought with him, and started slicing it. The school headmaster, Mr. Bedbrook, joined Percy and his guest, and offered tea and sherry to go along with the cake. Lamson happily accepted a glass of sherry, but remarked that he always took a little sugar with it to counteract the effects of the alcohol. No doubt Bedbrook thought this odd, but being the consummate host, the headmaster called for the matron to bring some sugar from the kitchen to sweeten Lamson's drink.

During the conversation Lamson broached the topic of a new kind of gelatin capsule that he had acquired, that could be filled with all manner of medications and would be perfect for giving bitter medicines to the pupils at the school. To prove his point, Lamson filled a capsule with some of the sugar, put the two halves together, and gave it to Percy, saying, "Here, Percy, you are a good pill taker, take this and show Mr. Bedbrook how easily it can be swallowed."[1] As soon as Percy had swallowed the pill, Lamson begged his leave, announcing that he had a train to catch.

As Bedbrook escorted Lamson to the school's entrance, Lamson noted that, in his professional opinion, he did not think Percy would live much longer. This was a considerable surprise to Bedbrook, who thought Percy looked quite healthy. Minutes after Lamson's departure, though, Percy began to complain of heartburn. He was carried to his bed, where his illness intensified. An hour later Percy was found lying on the floor beside his bed in great pain, and retching up a black fluid. His whole body convulsed, and he had to be forcibly restrained. Two doctors were called, but both were at a loss to explain the boy's symptoms; later they would confess to being ignorant of the effects of aconite on the human body. With little they could do, the doctors gave Percy a morphine injection to dull the pain, but at 11:30 p.m., after hours of torment and agony, Percy finally lost consciousness and died.

Although the doctors attending Percy were unfamiliar with

aconite, they were nonetheless convinced that he had been poisoned, since little else was known that could precipitate death so easily. Suspicion promptly fell on Lamson, who strongly protested his innocence. An autopsy of Percy's body was ordered, but there were no obvious signs as to what had caused the youth's demise. Convinced that a plant alkaloid had been employed, the police solicited the assistance of Dr. Thomas Stevenson, an expert in alkaloid poisons, who was based at London University. At the time, chemical tests for alkaloids were crude and not very sensitive, but Dr. Stevenson's party trick was his ability to detect and identify alkaloids by taste. During his career Stevenson had collected some eighty different alkaloids in his laboratory, and delighted in challenging his colleagues to identify the compounds through chemical tests faster than he could identify them by taste; Stevenson always won the challenges.

While detecting plant alkaloids by taste alone is an impressive, if strange, hobby, Stevenson was able to "taste" the alkaloids against a background of various body fluids, and so extracts from Percy's vomit, stomach contents, and urine were each in turn placed upon the cutting edge of nineteenth-century chemical analyzers: Stevenson's tongue. With all the expertise of an experienced sommelier, identifying not only the vintage of an unknown wine but also the field in which the grapes were grown, Stevenson started his "tasting." Extracts from Percy's stomach contents placed on Stevenson's tongue produced a "burning sensation, extending down to the stomach . . . peculiar to aconitia."[2] Stevenson's dedication to his testing method was such that he endured the symptoms for almost seven hours before they slowly wore off. Stevenson confirmed his conclusions by injecting a sample of Percy's urine under the skin of a mouse, with the resultant death of the mouse within thirty minutes. As a control, more mice were injected with a solution of prepared aconite, those mice dying with the same symptoms as those observed in mice injected with Percy's urine. The only conclusion that could now be reached was that Percy's death was the result of aconite poisoning.

Lamson was arrested and tried for Percy's murder at the Old Bailey in February 1882. Lamson, who entered a plea of not guilty was defended by Montagu Williams, who was quick to point out that none of the doctors or chemists involved had ever come across a case of aconite poisoning before, and therefore were in no position to definitively argue that Percy had indeed died from aconite. And regardless of whether aconite was the cause, nobody had actually seen Lamson administering a fatal dose of poison to his brother-in-law. The defense also attempted to cast doubt on Professor Stevenson's "tasting" of aconite in extracts from Percy's organs. Francesco Selmi, Professor of Medical Jurisprudence at the University of Bologna, had argued that poisonous alkaloids were naturally produced in the stomach of those who had died from natural causes, through a process of putrefaction. These cadaveric alkaloids, which Selmi called ptomaines, from the Greek word for "corpse," could easily be what Stevenson had detected.

The prosecution, led by Solicitor General Poland, brought Professor Stevenson back to testify, and he clearly impressed the jury as a master of his profession. When asked if cadaveric alkaloids could have been what he found in Percy's samples, he stated that the issue of cadaveric alkaloids was still a matter of dispute among experts, and that while some might mimic the effects of plant alkaloids, he was unaware of any that mimicked aconite. In any case Poland essentially destroyed the defense's notion of putrefaction alkaloids by pointing out that Percy's body had not yet started to decay when samples were taken and preserved.

Additional evidence against Lamson was brought by the pharmacist who sold him the aconite, and remembered the purchase clearly. At six o'clock on the last day of the trial, the jury deliberated for a mere thirty minutes before arriving at a guilty verdict. When Lamson was asked if he had anything to say in response to the verdict, he replied: "Merely to protest my innocence before God." The judge stated that "it would serve no good end were I to recapitulate the harrowing details of your cruel, base and treacherous crime. . . . I entreat you to prepare to meet Almighty God."

The execution by hanging was set for April 4, but before that date arrived, an intervention came from America. It was claimed that insanity ran in Lamson's family (a grandmother and other family members having been held at various times in Bloomingdale Asylum for the Insane in New York), and therefore Lamson could not be held entirely responsible for his crime. However, this news served only to delay the inevitable. No insanity plea had been entered during the trial, and so the sentence stood. Friday morning, April 28, 1882, was heavy and overcast at Wandsworth Prison. Lamson rose at his usual time to breakfast on coffee, eggs, and toast. At 8:45 a.m., as a light rain started to fall, Lamson was led to the scaffold.

It is certainly disconcerting that Dr. Lamson was more afraid of facing up to his debts and his morphine addiction than he was of murder and the hangman's noose. Lamson's father wrote a letter to the London newspapers stating that he would have been happy to cover all his son's debts, had he only asked. In prison Lamson was forced to give up his morphine addiction, and four days before his execution, in a moment of mental clarity, he confessed to Percy's murder. Finally, the question of why Lamson chose aconite as his weapon was revealed. A key piece of evidence at Lamson's trial was a notebook in which he had carefully written down the symptoms of aconite poisoning, and the fact that it was undetectable. That is, undetectable according to Edinburgh professor Robert Christison in the early 1870s, when a young medical student by the name of George Henry Lamson was diligently taking notes on the professor's description of aconite. Aconite was not to remain undetectable for long, and had Lamson kept up with the scientific literature, he might have chosen a different poison, or even abjured murder altogether.

THE ALKALOID PROBLEM

Alkaloids are naturally occurring organic compounds made from carbon, hydrogen, and a smattering of nitrogen, all woven together to make a molecule that has profound physiological effects on humans and other animals. The first major step in elucidating plant chemicals came from the German chemist Friedrich Sertürner in 1804, when he took opium poppies and purified from them the "soporific principle," which he called "morphium" in honor of Morpheus, the Greek god of dreams. The term *alkaloid* was introduced in 1819, and was based on the observation that when certain plant extracts were dissolved in water they made an alkaline solution. Between 1818 and 1860, several alkaloids were purified from plants, including strychnine (1818), quinine and caffeine (1820), nicotine (1828), atropine (1829), and cocaine (1860).

Although scientists were becoming more sophisticated in extracting and purifying alkaloids from plants, they had no idea how to detect those compounds in a dead body. Stevenson's taste test was available, but it required experience and expertise. It was quite subjective, and certainly could not determine how much poison was in a body. During one unsuccessful prosecution of a murder with morphine, a French prosecutor ranted in the courtroom: "Henceforth, let us tell would-be poisoners, use plant poisons. Fear nothing; your crime will go unpunished. There is no *corpus delecti* [physical evidence of a crime] for it cannot be found." Even the noted Spanish chemist Mathieu Orfila, the founder of toxicology and author of the first book on poisons and their detection, lamented that detecting alkaloids in a corpse might be an impossible task. This seemed to be a green light for poisoners, and a later analysis of poison prosecutions in nineteenth-century Britain showed that plant alkaloids were by far the most popular choice for committing murder.

In 1851 the first crack in the belief that alkaloids could not be detected came from a murder trial in Belgium, when Hippolyte Visart de Bocarmé poisoned his brother-in-law with nicotine. The Belgian chemist Jean Servais Stas, was asked to help with the prosecution's case by proving the presence of nicotine in the murder victim. Stas searched for three months for a way to isolate nicotine from human tissues, before finally discovering that treating tissue extracts with ether and chloroform would allow nicotine to be detected. Stas proved the toxicity of the nicotine he had extracted from the corpse by feeding small amounts to some pigeons and swallows. The birds went into convulsions and died within minutes.

The successful prosecution for murder by nicotine had three effects: First, the convicted Count Hippolyte was executed by guillotine (a spectacle watched by crowds in the thousands). Second, would-be murderers were put on notice that their poisons were not as undetectable as they might think. And third, the role Stas played in figuring out how to extract nicotine from dead bodies was honored by referring to the technique as the "Stas process." Despite this apparent success in nicotine detection, the large structural differences between molecules of nicotine and other alkaloids, such as aconite and strychnine, meant that no single extraction procedure would work for all alkaloids on all tissues. In fact, the question of whether plant alkaloids could be isolated and identified from cadavers would be a contentious point for many years. Determining that a person had died from alkaloid poisoning was as much a matter of knowing the victim's dying symptoms as performing chemical studies, and methods for detecting and analyzing alkaloids were not created until the middle of the twentieth century.

The problem of detecting aconite in a murder victim was highlighted in the case of John Hendrickson, tried for the murder of his wife in 1853 in New York. Hendrickson was the kind of man every father fears his daughter marrying. He had no job, was constantly drinking, and his wife's pregnancy sent him looking for the comfort of other women. Hendrickson's wife was not impressed with his activities and announced that she was going to divorce him and move back in with her recently widowed mother. But before she could put her plans into action, Maria was found dead in her husband's bed.

Suspicion arose immediately that this was not a natural death. Maria's body was placed in a coffin and taken to her mother's house, where, in the front room, a postmortem was conducted. The preliminary finding was death by poisoning. Sections of Maria's intestines were removed for analysis before the body was interred. Deciding that more tissue was needed for chemical analysis, the coroners dug up Maria's body five days after the funeral, and removed all her intestines.

Initial tests for arsenic, cyanide, and a few other poisons all came up negative, leaving the police stumped as to what the poison might be. A thorough search through numerous medical textbooks suggested that the mystery poison might be aconite. But was there a test for aconite? Certainly the Stas process could be used to detect aconite's chemical cousin, nicotine, but there was no evidence that it could be used to find aconite as well. Indeed, the Paris College of Pharmacy had put up a considerable amount of prize money for anyone who could devise a test for aconite; the prize remained unclaimed.

This did not deter the prosecution's star witness, Dr. James H. Salisbury, an expert medical chemist and accomplished medical witness in poisoning cases, from feeding the victim's stomach contents to a cat and concluding that he had definitive evidence of aconite poisoning. When the defense asked for a sample that they could put to their own tests, the prosecution's expert claimed he had none left. The prosecution expert claimed that he had "tasted it, tested it, and satisfied himself that he had detected aconite."

When asked by the defense how the cat fared, Salisbury admitted that the cat did not die and was alive and unharmed. The court transcripts provide a wonderful nineteenth-century oration by the defense:

> Just look at it—the confidence of this Dr. Salisbury. He, so he says, has discovered this aconitine; he has solved the great problem; and yet calls no one in to see his discovery, or to confirm it. He is in too great a hurry; he cannot wait; but administers it all to a cat. He could not wait, he had such a desire to send his name abroad; he could not stop a single moment, could not bring a particle of it into court for us to see it, or taste it; but he gives it all to a cat!
>
> . . . .

> The cat did not vomit, retained it all, and in three hours was well. What a cat! What a doctor! What an opinion founded upon such facts! The cat should have died out of deference to the Dr.'s opinion, or the Dr. should have given up his opinion out of deference to the life of the cat.[3]

Although questions were raised as to whether the expert witnesses had really found aconite, Hendrickson was hanged in the courtyard of the Albany County Jail on May 5, 1854, still desperately claiming his innocence.

HOW ACONITE KILLS

Soon after ingesting aconite, the stomach and intestines become unsettled, as the body tries to rid itself of the deadly poison. Nausea, vomiting, stomach cramps, and diarrhea all occur quickly in an attempt to physically remove the aconite. Usually these attempts are in vain, as some of the poison has already been absorbed into the bloodstream. Carried throughout the body in the blood, aconite's first deadly symptoms appear. Starting with a feeling of pins and needles around the mouth, numbness gradually spreads over the entire face. Often a burning sensation is felt on the tongue, as if a red-hot poker is being slowly drawn over it. The eyes lose focus, and vision begins to blur and fade, even to the point of blindness. The normal sensations of the arms and legs disappear, almost as if they have been cut off from the rest of the body.[4]

As aconite works its way through the body, the skin grows cold and clammy. Labored breathing and a sense of dread overwhelm the victim. Blood returning to the heart carries the poison with it, initially causing palpitations, but then causing the heart to beat faster and faster until the heartbeat becomes erratic and finally ceases altogether. The effects of aconite poisoning are seen quickly, usually within minutes of ingestion, and seldom delayed longer than an hour. Once a lethal dose has been ingested, the only remaining question is whether death will result from paralysis of the heart, or asphyxiation due to paralysis of the diaphragm. As little as 1 or 2 milligrams, an amount roughly equivalent to 100 to 200 grains of salt, is fatal. Even with hospital support, 95 percent of patients will succumb and die. This is a very serious poison, and is aptly named "the queen of poisons."

Aconite binds to a particular protein found in the membranes of nerve and heart cells. Both nerve and heart cells need small electrical currents to work properly, and aconite wreaks its havoc by interfering with this bioelectricity. In contrast to the continuous flow of electricity down a wire, nerves work by sending waves of signal down their length. Once a signal has washed down the nerve, the nerve must reset before it can again conduct a signal. Similarly, after each heartbeat, the heart must momentarily rest and reset before starting another contraction to send blood around the body. If either the nerves or the heart are unable to reset properly, problems quickly follow. It is this resetting process that aconite prevents from happening.

When a nerve fires, its sodium and potassium ions swap places. Sodium ions, which are normally sparse inside the cell, flood into the nerve, prompting potassium ions to exit the nerve. This exchange of sodium going in and potassium leaving is called depolarization, and it controls nerve signaling. Sodium doesn't just leak into the nerve; it enters in a carefully controlled fashion through specialized proteins in the membrane called sodium channels. To reset the nerve so that a new signal can be sent, a process of repolarization must happen, in which the sodium channel closes to stop the inward flow of sodium into the cell, and any sodium that has come in must be expelled. In the muscle cells of the heart, the influx of sodium triggers the contraction of the muscle, and when all the muscle cells of the heart contract in a coordinated fashion, we have a heartbeat. After each contraction of the heart, the cells must repolarize, and the sodium channels must shut down.

Imagine now that something prevents these sodium channels from ever closing. That is what aconite does. Aconite stops the nerves and heart muscle cells from repolarizing and resetting by binding tightly to the sodium channels, like a doorstop blocking a door from shutting. At first the sodium channels open, and sodium floods into the cell, causing a nerve signal to fire or a heart cell to contract as usual. After a few milliseconds the sodium channel should close to reset the system, but aconite locks the channel open. Nerve and muscle cells try to get the sodium ions back out using sodium pumps (we will come across these again), but with the channels wide open, it's like trying to empty a bathtub with the faucet still running.

If aconite is so deadly, why would it ever be used medicinally? Not all nerves send information from the brain out to the body.

Some nerves are sensory, and send information from our senses back to the brain; this group includes pain nerves. Although these nerves are incredibly useful in helping prevent damage to the body, long-term pain is unpleasant. Since sensory nerves also rely on sodium flow and depolarization, preventing the closure of sodium channels in these nerves should eliminate the pain signals. This is the basis for herbal pain remedies based on aconite. Unfortunately, the tiny amount of aconite needed to curb pain sensations is perilously close to lethal levels. Several instances of aconite poisoning have been attributed to the use of herbal remedies that undergo no quality control or purity assessment.

ACONITE AND MRS. SINGH'S CURRY

After George Lamson's execution, aconite and its poisonous properties slowly faded from the public's mind, but the echoes of "murder" and "aconite" were to reverberate through the chambers of the Old Bailey courthouse again, almost 130 years later.

Lakhvir Kaur Singh was born in Amritsar, India, before moving to the London borough of Southall. In an arranged and loveless marriage and with three children, she felt trapped. Mrs. Singh decided that what she needed was some excitement in her life. This soon arrived in the form of Mr. Lakhvinder "Lucky" Cheema, a relation of the Singhs by marriage, who moved into their home as a lodger and eventually became Mrs. Singh's lover. Spending more than a second in thought, one would realize that going by the nickname "Lucky" was just tempting fate. After living with the Singhs for several years, Lucky finally moved out, got a house of his own, and started taking in lodgers to help pay the bills. This had little effect on Mrs. Singh, as she continued to act the part of a devoted mistress by visiting Lucky's house every day to clean, cook, and do his laundry.

The relationship began to unravel in 2008 when Lucky Cheema was introduced to twenty-one-year-old Gurjeet Kaur Choough, an immigrant to the UK. Choough was introduced to Lucky as a potential bride; from that point on, Lucky started to run out of luck. Barely a month later, Lucky and Choough announced their engagement, carefully timed for when Mrs. Singh was in India visiting relatives.

Despite being thousands of miles away, Singh learned of the engagement. Furious and distraught, she bombarded Lucky with text messages, begging him to return to her. In one text, Singh wrote: "Did you not think before breaking my heart that my heart would now be useless to anyone?" She also tried to persuade Lucky that his young fiancée only wanted to marry him to secure legal residency in Britain. When none of these approaches worked, Singh decided that if she couldn't have Lucky, then no one would. One visit to an herbal remedy store later, Singh left with a packet of *Aconitum ferox*, or Indian aconite powder, which she smuggled back into England.

Back home, Mrs. Singh kept a close watch on Lucky's house, noting when he left and when he returned. With this knowledge in hand, on January 27, 2009, Mrs. Singh patiently waited until Lucky left the house and drove off. Using the keys Lucky had given her long ago, Mrs. Singh entered the house, briefly waving to one of Lucky's lodgers, and went directly into the kitchen. Inside the refrigerator was a large Tupperware container holding some chicken curry. Reaching into the fridge, she opened the plastic tub and carefully sprinkled in deadly aconite.

When Lucky returned home, the lodger told him that Mrs. Singh had visited while he was out. Lucky thanked him for the information, determining that he really needed to change the locks on the front door in case Mrs. Singh ever decided to perpetrate any retribution. If one were to look for a picture to go along with the phrase "locking the stable door after the horse has bolted," a photograph of Lucky Cheema and a Tupperware bowl of curry would not be inappropriate. At 10:00 p.m. that night, Lucky and his fiancée sat down for a late dinner, tucking into the warmed curry from the fridge. Discussing their upcoming wedding, which was to take place in two weeks, on Valentine's Day, Lucky took a second helping of the food. Not much later both Lucky and Gurjeet felt unwell, complaining of severe stomach cramps. Lucky called emergency services and, in a trembling voice, told the operator that he thought someone had poisoned their food. Deciding that the ambulance would take too long to arrive, Lucky got his two nephews to take him and Gurjeet to the hospital.

Partially paralyzed and losing his sight, Lucky and his fiancée were helped into his car and taken to the emergency room. The initial symptoms recorded by the doctors included a feeling of pins and needles around the mouth, loss of vision, muscle weakness, sweating, abdominal pain, and profuse vomiting. Despite being treated with antiemetics, both patients continued vomiting. Within an hour of admission, Lucky became very agitated and his heart started racing. Machines monitoring Lucky's heart showed massive changes in electrical activity, causing it to contract erratically, and—importantly—inefficiently. Lucky's blood pressure plummeted, he started convulsing, and within two hours of admission, he was dead.

The doctors had no clue what had poisoned Lucky, but started taking steps to wash out any drug from Gurjeet's stomach. She was placed in a medically induced coma for three days, and likely survived only because she'd eaten less of the curry than her late fiancé.[5] The rapid onset and fatality of the poison alarmed the hospital staff and police, who sealed off Mr. Cheema's and Mrs. Singh's homes, suspecting an airborne or chemical threat. In Singh's coat police found a plastic bag containing a brown powder, which she claimed was just an herbal remedy.

At first the forensic chemists were unsure of the identity of the powder, but whatever it was, the chemical was also present in the curry, and Lucky's vomit. Suspicion fell on an alkaloid from *Aconitum ferox*, which grows in the Himalayas, but the forensic chemists couldn't just fly off to the Himalayas for samples to see if their hunch was correct. Fortunately, samples of *A. ferox* were available much closer to home, in the Royal Botanical Gardens at Kew. When the samples from Kew Gardens were compared with Mrs. Singh's brown powder, they were found to be identical. Lucky's death was due to poisoning with aconite from *Aconitum ferox.*

Lakhvir Singh was tried at the Old Bailey for the murder of Lucky Cheema by poisoning, and the attempted murder of Gurjeet Choough. Great public interest surrounded the trial, since the last trial in Britain for aconite poisoning had been 130 years earlier when George Henry Lamson had been convicted in 1882. The jury found Mrs. Singh guilty of the murder of Lucky Cheema, and guilty of causing Ms. Choough grievous bodily harm with intent. In sentencing, the judge said: "You set about a cold and calculating revenge. You knew how deadly aconite was and how agonizing the effects would be." Singh received a life sentence with a twenty-three-year minimum term. Some 130 years apart, two murderers, both using aconite, and both tried in the same courthouse, were each found guilty. Fortunately for Mrs. Singh, by the time she was convicted, hanging was no longer a sentencing option.

In the deadly cat-and-mouse game between poisoner and toxicologist, the poisoner certainly had the upper hand at the beginning of the nineteenth century. Poisons, including plant poisons, were readily available, and even if someone was suspected of murder by poison, forensic evidence as a prosecutorial tool was in its infancy. However, as the twentieth century rose on the horizon, chemists and toxicologists became more sophisticated in their abilities, to the point where killers who might have gotten away with murder a few years before were finding their poisons identified in their victims. Today's modern toxicology labs, filled with cutting-edge detection equipment, mean that no substance is ultimately undetectable. In the next chapter we look at an equally lethal plant poison, but one that kills in a completely different way: by interfering with a critical activity in every one of the thirty trillion cells in our bodies.

. { 5 } .

Ricin and Georgi's Waterloo Sunset

Poison has a certain appeal. . . . It has not the crudeness of the revolver bullet, or the blunt instrument.

—Agatha Christie, *They Do It with Mirrors*, 1952

LABORATORY NUMBER 1

In the movies—and in real life—spy agencies are always trying to devise new and undetectable ways of eliminating their opponents. Among the most notorious of these was the Soviet Union's KGB (Committee for State Security), now known as the Russian Federation's FSB, or Federal Security Service. Both agencies adopted an uncompromising policy of eliminating anyone deemed a threat to security, and a key element of these assassinations was to make the death appear to be of natural causes. The development and production of specialized poisons that were extremely hard to detect, identify, and trace was the function of a top-secret research laboratory in Moscow, called Laboratory Number 1, housed on Varsonofevsky Lane near the KGB's Lubyanka headquarters.

When the head of Laboratory Number 1 once complained that poisons tested on animals were not always effective in humans, Lavrenty Beria, Stalin's head of security, smiled and, with all the ominous enthusiasm of a James Bond villain, asked, "Who's stopping you from doing experiments on humans?"

The trademark of Laboratory Number 1 was taking existing poisons and using them in a way that was difficult to detect or trace back to Russia. It is not possible to know how many undetected assassinations were carried out by poisons developed in the lab, but some have become notorious. In 1957 Munich, anti-Soviet agitator Lev Rebet was sprayed in the face with cyanide mist, squirted from a spray gun hidden in a rolled-up newspaper. So successful was this assassination that it was determined that Rebet had died of natural causes from a heart attack. It was only four years later, when the assassin defected to the West and revealed the plot, that Rebet's death was deemed a murder. Nikolai Khokhlov, a KGB defector, drank a contaminated cup of coffee at a public reception in Germany in 1957, falling gravely ill shortly thereafter. Blood tests revealed the presence of thallium, a metallic toxin used in rat poison, but all attempts at treatment had little effect. When Khokhlov was transferred to a U.S. Army hospital in Frankfurt, doctors discovered that the thallium had been exposed to radioactivity so that the metal would slowly disintegrate in his body, giving the appearance of severe gastroenteritis as the patient slowly died. Soviet defector Georgi Okolovich narrowly escaped assassination when an attempt to shoot him with a poison-tipped bullet, fired from a miniature pistol concealed in a cigarette pack, failed.

A miniature spray gun, a deadly coffee cup, and a tiny pistol in a cigarette pack were just a few of the bizarre delivery systems created by scientists at Laboratory Number 1. However, perhaps the most notorious of all the devices conceived to bring about death was an umbrella, created for the Bulgarian Secret Service and designed to eliminate dissident Georgi Markov.

THE STORY OF THE CASTOR BEAN

A spoonful of castor oil was seen as a panacea for many common childhood ailments, and still retains a reputation as a home remedy. However, the plant that yields oil that's safely ingested by thousands of reluctant children is the same plant that produces the most dangerous toxin known to man.

Despite its notoriously bad taste, castor oil is a safe and relatively mild laxative, sold as an over-the-counter medication. In fact, it is not the oil that has a bad taste, but rather the result of a reaction between the oil and air. The use of castor oil as punishment was perfected in Fascist Italy under Benito Mussolini, where it was a favored tool to humiliate opponents. Mussolini's Black Shirts would force-feed large quantities of castor oil to political dissidents. Victims of this treatment did sometimes die from dehydration, though the nightstick beating that often accompanied the administration of the oil likely also had some effect. It was once thought that castor oil worked by irritating the lining of the intestine, causing inflammation in the gut, but it is now known that castor oil locks onto specific receptors, increasing the contractions of the smooth muscle cells of the intestine.

Chemically, castor oil is made from ricinoleic acid, and has tremendous commercial value, with a wide range of uses from the preservation of wood and leather to the manufacture of brake fluid, paints, and inks, and as a lubricant for heavy machinery. The other product of the castor plant is ricin, a substance that not only lacks any of the commercial or medical benefits of the oil, but is deadly even in small amounts.

Castor plants are large, robust shrubs that can grow from six to fifteen feet in a single season. The shiny seeds of the castor plant are called beans and feature very beautiful, intricate designs. Deadly ricin is found in small amounts throughout the castor plant, but is present mostly in the beans. Specifically, ricin, along with castor oil, is found in the endosperm of the castor seed, the source of nutrition for germinating seedlings. As with all plant alkaloids, ricin may serve as a deterrent to animals' eating the seeds and young plants (although birds such as ducks, hens, and doves are relatively immune to ricin poisoning). Castor oil is extracted by heating the seeds to 284° Fahrenheit (equivalent to 140° C) for about twenty minutes, to break down and inactivate the ricin protein, followed by crushing and pressing the beans to release the oil. The remaining husks can be used as fertilizer, but not as livestock feed, since they still contain trace amounts of ricin.

Harvesting castor beans is not without its dangers—not only due to the presence of ricin, but also the plant's large production of pollen, which contains allergens that can induce asthma. The sap, flowers, and leaves can cause a painful skin rash on contact. Prolonged exposure to these allergens can also lead to permanent nerve damage. For these reasons many companies are trying to find alternative sources of ricinoleic acid, or, through genetic engineering of the castor plant, to generate a plant that yields the important oil but is devoid of ricin or allergens.

THE TRUTH THAT KILLED

Georgi Ivanov Markov was born in Sofia, the capital of Bulgaria, on March 1, 1929. As a teenager, Markov saw his country become a one-party socialist state, and by the mid-1960s, President Todor Zhivkov had turned Bulgaria into one of the Soviet Union's staunchest allies and one of the most repressive regimes in the Warsaw Pact nations. Markov grew up to become an acclaimed novelist and playwright, earning the prestigious Union of Bulgarian Writers annual award for his first novel, *Men*, in 1962. His work found favor with the Communist administration, and he moved freely among the upper echelons of the social elite and political grandees of the Communist Party leadership.

Despite the privileged life he led in Bulgaria, Markov gradually became disillusioned with the glaring corruption and repression of freedom in his country, and in 1969 he started rehearsing a new play in secret, a play that was not favorable to the Communist leadership. Following the first performance, Markov was summoned to face the Committee on Culture to answer for his anti-Communist propaganda. Wisely, Markov refused to go, and escaped to the West. In his absence he was tried and convicted of being a traitor to the Bulgarian state.

Markov fled first to Italy, where he stayed with his brother for a short while, before finally ending up in London to start a new life as a journalist and writer. In 1975 he became a broadcaster on the CIA-sponsored Radio Free Europe network. In his weekly shows Markov expressed strong anti-Communist views and used his literary skills to expose the corruption in the upper ranks of the Bulgarian government, gaining him a large audience with the people of his native country. Understandably Zhivkov's regime did not appreciate such condemnation of Bulgaria's lack of human rights and their suppression of democracy. When the Bulgarian authorities refused permission for Markov to return to visit his dying father, his broadcasts veered into a series of personal attacks on President Zhivkov. The Bulgarian authorities naturally took a very dim view of Markov's assertions, and by June 1978, the Bulgarian government decided to embrace a classic line from *The Godfather*, and make Markov "an offer he couldn't refuse."

The offer was simple: Stop broadcasting for Radio Free Europe or be executed. Markov declined the offer, and the decision was made that he had to be silenced. The Darzhavna Sigurnost, the Bulgarian Security Service, sought advice from their Soviet big brother at Laboratory Number 1, as to the best way to eliminate their problem. What ensued was one of the strangest killings of the Cold War. After Markov's death, scripts of his radio broadcasts were collected and published in a book called *The Truth That Killed*.

THE ASSASSINATION OF GEORGI MARKOV

At the end of August 1978, Markov received a strange and disturbing phone call. The caller informed Markov that he would soon die of what looked like natural causes, but would really be something out of the ordinary. Two weeks later, Thursday, September 7, was President Zhivkov's birthday, and the Bulgarian Secret Service planned on giving their leader a special birthday present: Georgi Markov's death.

Markov followed his usual routine that day, and around noon parked his car near Waterloo Station, on the South Bank of the Thames. Markov left the parking lot and walked the short distance to the bus stop, where he waited for a bus to take him to work on his weekly radio broadcast. As he was waiting, Markov felt a sharp sting on the back of his right thigh and spun around to see what had caused it. He appeared to have been prodded accidentally with an umbrella, since a man nearby was bending down to pick up a rolled umbrella that had fallen on the ground. The man apologized to Markov in a foreign accent before hailing a taxi and leaving.

Markov caught the bus to the BBC World Service offices, and finished his broadcast. As the afternoon and evening wore on, he started feeling a little out of sorts, as if he were coming down with a cold. By the time he arrived home that night he was not feeling any better, and so as not to disturb his wife, or pass on his cold, Markov made up a bed in his study. At 2:00 a.m., Markov's wife, Annabel, was awoken by the retching sounds of her husband's violent vomiting. His temperature had soared to

104° F (40° C), and—concerned for her husband's well-being—telephoned a doctor for assistance. Given the symptoms, the doctor determined that Markov was coming down with the flu, and that he should rest in bed and drink plenty of fluids. The doctor had no way of knowing that Markov was now part of an international assassination plot, and that his body was slowly shutting down in response to a deadly poison.

The next day Markov continued to deteriorate, and by the afternoon he was struggling to speak. On the evening of September 8, 1978, Markov was admitted to St. James's hospital in the Balham neighborhood in south London. In the emergency room, surrounded by the usual assortment of accidents, cuts, heart attacks, and stomach pains, was a man who claimed he'd been shot by the KGB.

Startled by the patient's bizarre claims, the house physician, Dr. Bernard Riley, listened as Markov explained that he was a Bulgarian defector who had enemies back in Bulgaria, and that his friends had warned him that the KGB were "out to get him." To Riley, the man in the ER bed seemed paranoid or delusional. Certainly the patient was feverish and had an elevated temperature, but this could have been from any standard infection, like influenza or gastroenteritis. Markov also complained of nausea and vomiting. Recalling the previous day's events, Markov was convinced he had been shot in the leg with a poison dart, but although a thorough examination of his thigh revealed inflammation around a small puncture wound, X-rays failed to reveal the presence of any foreign body.[1]

By the evening of Saturday, September 9, Markov was in a state of steady decline, and had been transferred to intensive care. His blood pressure, now 70/40 compared with the normal 120/80, was dangerously low, with his heart racing at 160 beats per minute. Even though he was sweating profusely, Markov complained of feeling cold. Samples of blood were taken, and Markov's white cell count (the blood cells responsible for fighting infections) were extremely high at 27,000 (the normal range being 5,000–10,000). All these symptoms suggested septic shock and widespread infection, but though Markov was pumped full of antibiotics, he failed to respond. He continued vomiting, though now his vomit was flecked with blood, a sign of internal hemorrhaging as the lining of his stomach and intestines slowly started to fall apart. As the day wore on, Markov also stopped passing urine, a sign that his kidneys were starting to shut down.

As Markov's kidneys stopped working, fluid began filling the space around his lungs, making breathing labored and inefficient. The next day an electrocardiogram was taken and showed that Markov's heart, too, was failing, showing erratic irregular beating. Early Monday morning Markov became confused and delirious and started pulling out his intravenous lines. At 9:45 a.m. he went into cardiac arrest, and despite vigorous attempts at resuscitation, at 10:40 a.m. on September 11, a mere four days after the incident on Waterloo Bridge, Georgi Markov was pronounced dead. He was just forty-nine years old.

DEATH IN A PELLET

Because Markov was a well-known dissident, and had recently received death threats, the police and Scotland Yard took his assertions of an assassination much more seriously than the doctors had. To find out if Markov had really been poisoned, an autopsy was ordered. Dr. Rufus Crompton, the Home Office pathologist working for the British government, found massive damage to Markov's heart, lungs, liver, intestine, and pancreas, as well as substantial hemorrhaging in other organs. The lymph glands, particularly those in the groin area on the right side—the side Markov had complained of being stung on—were swollen, suggesting something had gone into the back of Markov's leg, and then traveled up to his lymph glands before entering his circulation. The massive swelling of the lymph glands indicated that Markov's body was fighting some kind of toxin.

Scotland Yard decided that further investigation required specialists, and called in scientists from Porton Down laboratories, the Ministry of Defence's top research station for biological weapons during the Cold War. Carefully going over Markov's body, inch by inch, medical officer Dr. Robert Gall noticed a small ball bearing–shaped object buried in Markov's thigh. The object was a small metallic bead, with what appeared to be two holes drilled in its middle. The pellet was made from an iridium-platinum alloy, probably chosen because it could escape detection by the body's immune system. The two holes allowed it to act as a reservoir for some toxic agent, and the pellet was likely covered with a gelatinous coating to keep the poison inside. A quick calculation suggested that the inside of the pellet could hold around 400 nanoliters of fluid (4 billionths of a liter) or 500 micrograms (500 millionths of a gram) of material, but the pellet was empty and provided no indication of its lethal contents.

Tests for bacteria in Markov's blood were negative, suggesting the pellet had not housed bacteria. What about a virus? Certainly the pellet was big enough to contain many virus particles packed into its interior, but Markov's death was too sudden and his symptoms too rapid in onset to have been caused by viruses. So, what about bacterial toxins? The likely candidates—diphtheria toxin or tetanus toxin—didn't give rise to the same symptoms that were experienced by Markov, and in any case, most people had received immunizations against those. Chemical poisons like arsenic or cyanide were considered, but though they are extremely dangerous, the small amount that would have been housed in the pellet was at least ten times smaller than a lethal dose of cyanide. As stated by the great Sherlock Holmes, "Once you eliminate the impossible, whatever remains, no matter how improbable, must be the truth." What remained now were natural plant toxins. Of these, ricin from the castor oil plant seemed the most likely. But could the pellet have contained enough ricin to kill a man?

Scientists had some information on the effects of accidental castor bean poisoning, but this was the first time that someone had been deliberately injected with purified concentrated ricin. To determine if it was possible that Markov had been poisoned by ricin, the researchers decided to test the effects of a small amount of pure ricin on an animal. Pigs are about the same weight as a man and have a remarkably similar circulatory system. A pig was injected with the same amount of ricin as it was thought Markov had received. For the first six hours, nothing happened—and then the pig became extremely sick. It developed a fever, and had a raised white cell count, just like Markov. The pig was obviously very ill, and an ECG showed a severely abnormal heart rhythm, just like Markov. In a little over twenty-four hours, the pig was dead. An examination of the pig revealed the same internal damage as seen in Markov. Although it was impossible to prove definitively that ricin was the deadly agent in the pellet, the coroner's office ruled that, based on the preponderance of evidence, Markov had been killed with 450 micrograms of ricin toxin.

Piecing together the likely chain of events, it was presumed that the capsule had been coated with wax to prevent the contents from leaking. Once inside Markov, the coating was slowly melted by his body temperature, gradually releasing the ricin into his bloodstream. But how did the pellet get into Markov's leg in the first place? The pellet was not deformed, as a bullet would be on entering the body, and there were no powder burns on Markov's jeans, suggesting that a conventional firearm was not the delivery weapon. The conclusion was that some kind of compressed-air or spring-loaded device had been used, but how was it concealed?

It is often said that the best place to hide something is in plain sight, and one thing that is certainly ubiquitous in London is an umbrella. Markov's own account of the days preceding his hospital admission strongly suggested that the umbrella that had poked him at the bus stop was, in fact, a compressed-air gun that had been triggered by the assailant as he pressed the umbrella tip against Markov's leg, shooting a pellet through Markov's pants and into his thigh. The use of a deadly toxin and the unusual delivery system pointed the finger of suspicion directly at Laboratory Number 1.[2]

HOW RICIN KILLS

Amazingly, an amount of ricin powder no bigger than a few grains of table salt is enough to kill. In contrast to the earlier poisons in this book, which work on the outside of nerve cells by interacting with specific protein molecules embedded in the cell membrane, ricin attacks every cell in the body—but it must first get inside the cells before it can unleash its destructive power.

Ricin takes us on a journey into the very heart of the cell, deep down to the cellular factories where the proteins necessary for life are manufactured. Our bodies need proteins to function, from growing hair and fingernails to producing the enzymes in our intestines to digest food. Proteins make up the cells of nerves, critical to conveying messages throughout the body, and the muscles of the heart that move oxygen to the body and brain. Still other proteins make up antibodies that protect the body from foreign pathogens.

Just as a sentence is made by stringing together a sequence of letters in a particular order, so proteins are made by stringing together amino acids in a specific sequence. In a sentence, each position can be held by any one of twenty-six letters, but with proteins, each position is limited to one of twenty amino acids. Not all sequences of letters can make a coherent sentence, and not all possible sequences of amino acids give rise to proteins. It is estimated that there are about one hundred thousand different unique proteins in humans, arising from one hundred thousand unique sequences of amino acids. Most people are familiar with the general idea that DNA houses the blueprint for who we are, and DNA does that by determining the order of amino acids that go into all our proteins.

When certain proteins are needed by the cell, copies of specific parts of that DNA blueprint are made in the nucleus, in a process called transcription. These copies, called messenger RNA, are then converted into proteins in a process called translation. Key to the translation process is a specialized complex in the cell called a ribosome—a large complex of proteins and nucleic acids that reads the genetic code and then uses it to link amino acids together in the right order. Like machines, proteins in the cells of the body are made and eventually wear out, being replaced with freshly made proteins. Some proteins only last a few hours, whereas others can last days; but whatever the protein, it will eventually wear out and need replacing. If the machinery replacing the proteins no longer works, the cell will gradually wear down, fall into disrepair, and eventually die.

Appropriately, ricin is classed as an RIP, or ribosomal inhibiting protein, since it destroys the ribosome, making it unable to manufacture new proteins. Ricin is made from two protein chains (named, with typical scientific flair, A and B), held together by a single chemical bond between two sulfur atoms. The two components of the ricin molecule act like a letter bomb: One part is the address that gets the bomb to its destination, and the other is the explosive that goes off when it arrives. The B chain of ricin can bind tightly to proteins embedded in all cell membranes, whereupon the A chain—the bomb—can inflict its damage. Hijacking the cell's normal processes, ricin is absorbed into the interior of the cell, where it seeks out the site of protein translation. Here the two ricin chains are broken apart, releasing the A chain. Once the A chain is on its own, it seeks out ribosomes and targets them for destruction. Unlike a bullet that can only be used once, the ricin A chain causes a chemical breakdown in one ribosome molecule, and then continues to roam around the cell seeking other ribosomes to destroy. In this way, one molecule of a ricin A chain can destroy fifteen hundred to two thousand ribosomes every minute.

At this rate it doesn't take long to kill every ribosome in the cell. Since the ribosomes are now inactivated, they are no longer capable of making new proteins either, and so, very rapidly, the cells of the body begin to break down. As ricin can destroy one ribosome after another, a single ricin molecule is sufficient to kill a cell. When enough cells are dead, lesions and hemorrhaging occur as tissues break down and blood enters the intestines and urine. More cell death, and there are no longer enough remaining live cells in the organs of the liver, heart, kidneys, and even brain to function.

The body does not go down without a fight, however. The presence of a foreign intruder is sensed by the body's immune system, which marshals its white cells to attack and kill the ricin molecules, while generating antibodies to destroy the ricin invader. This leads to a marked increase in the number of white cells circulating in the blood. However, the ability of just one molecule of ricin A chain to do so much damage means that an incredibly small amount of the toxin is extremely lethal. In theory, three micrograms (three millionths of a gram) of ricin contain enough ricin molecules to kill every cell in the body.

The specific symptoms experienced upon exposure to ricin depend on the delivery method. Inhalation of ricin causes irritation and bleeding in the airways and lungs. As more damage ensues, lethargy and fever follow. Blood and fluid seep into the lungs, making it harder and harder to breathe, eventually leading to death from respiratory failure. Injection of ricin causes some localized damage at the injection site, but as the toxin is carried through the body, fever, nausea, and hemorrhaging occur, leading to organ failure and death. Ingestion causes vomiting and nausea, along with gastric and intestinal bleeding and shock; death typically follows three to five days after exposure. Ingestion of ricin is slightly less dangerous, however, as the digestive system can break down a good deal of the ricin protein, inactivating it. An assassin would need to put one hundred times more ricin in food to kill a victim than if injecting it. Curiously, many plants contain the deadly ricin A chain, including barley and wheat. However, without the linked B chain to bind to the protein receptors, the A chain cannot get inside cells to wreak its havoc, so wheat and barley are safe to eat.

The attractiveness of ricin to assassins lies not only in the small amount required to bring about death but also because there is no known antidote or therapy. All that can be done is to treat the symptoms and provide pain relief to the dying victim. More sensitive methods to detect ricin have been developed, and clinical trials for vaccines have been undertaken. However, vaccines are used primarily for prevention, and have no value once ricin has taken hold in a victim.

WHO KILLED GEORGI MARKOV?

The assassination of Markov strongly pointed to the Bulgarians, with assistance from the KGB. The evidence certainly was circumstantial, but when taken together appeared very strong. What was lacking was definitive proof, but that proof was locked away behind the Iron Curtain. By the end of 1979 the investigation into Markov's murder had ground to a halt.

For ten years there was no new information; then came the collapse of the Eastern Bloc and the fall of the Berlin Wall. Many of the Bulgarian Security Service's files were destroyed by fire, likely by former members of the service trying to hide their actions during the Communist regime. But eventually files were unearthed that not only documented Bulgaria's involvement in Markov's assassination but also identified the person chosen to carry out the killing. The agent, code-named Piccadilly, was based in Denmark, and had received "special training" by Bulgarian Security to neutralize Markov. Upon completion of the mission, Piccadilly received two medals, several free vacations, and thirty thousand dollars.

But who was Agent Piccadilly? Twenty-seven years after

Markov's assassination, he was identified as Francesco Gullino, a Danish national using his business as an antiques dealer for cover. Gullino made several trips between Denmark and Britain in 1978, and according to Bulgarian Security files, was their only agent in London when Markov was "neutralized." Gullino left London the day after the attack and flew to Rome, where, it is alleged, he stood in a particular spot in Saint Peter's Square as a signal to his Bulgarian handler that the job had been done.

Although Gullino was later arrested and interrogated by Danish, British, and Bulgarian police, he was let go due to lack of concrete evidence. The circumstantial evidence against him is strong, but he denied any involvement in the Markov case and maintained that the insistence of Markov's death by assassination was part of an elaborate Cold War conspiracy against Bulgaria, and anti-Communism propaganda.

A DEADLY RETIREMENT PLAN

Little could be further removed from espionage and political assassinations at the height of the Cold War than a retirement home set in the bucolic environs of Shelburne, Vermont, close to Lake Champlain. Although the people and motives involved were quite different, the choice of poison was identical: ricin.

The upscale Wake Robin Retirement Home advertised its residents as "a vibrant engaged people, and a community in which you can be yourself." While most residents were content to fill their time with idle conversations or visits from family, seventy-year-old white-haired Betty Miller decided to take up a new hobby: Betty passed her retirement time experimenting with homemade poisons in her kitchen.

Above the stove in her small apartment she kept glass jars carefully labeled with their contents: "Cherry Seed," "Apple Seed," "Yew Seed," "Castor Beans," and "Ricin," quite an unusual collection for a retiree. More disconcerting was the presence of several pill bottles hidden in a wicker basket in the kitchen; the one labeled "Ricin" and containing a yellowish-white powder was already half-empty. Betty Miller had indeed been fairly industrious, scouring the internet for instructions on how to prepare ricin. Police would later find instructions labeled "How to make Ricin" that had apparently been printed off the internet using one of the residence's laptops.

Betty made her toxin by harvesting thirty to forty castor beans from plants growing on Wake Robin's property. From these she made about two to three tablespoons of ricin in her kitchen. In November 2017 Ms. Miller confided to the health care providers at Wake Robin that she had attempted to poison several of her fellow residents using her homemade ricin. She admitted to placing multiple servings of ricin in other residents' food and drinks over the previous few weeks. Staff at the retirement home immediately contacted the police, who also alerted the FBI, due to the hazardous nature of the toxin.

Within hours, the police, the FBI, Vermont's Hazardous Materials Response Team, and the National Guard's Fifteenth Civil Support Team descended on Miller's apartment. Remarkably, Miller was allowed to drive herself to the University of Vermont

Medical Center for evaluation and observation. The next day she was interviewed by the FBI, and told investigators that she had become interested in plant-based poisons the previous summer. When asked why she had made the ricin and given it to other residents, Miller explained that her goal was to commit suicide using the ricin, but she wanted to test its effects on others first.

Fortunately for the residents concerned, ingestion of ricin is the least effective route of administration. Indeed, none of the residents showed any signs of ricin poisoning, though one resident tested positive for ricin exposure. Despite Miller's review of internet recipes, she did not have access to the proper laboratory equipment needed to make purified ricin, and this, along with the breakdown of ricin in the stomach and intestines, likely saved several residents of the retirement home from serious illness. When she saw that none of her victims appeared any the worse as a result of her experiments, she made a stronger concoction and put it in a "friend's" cup of tea. Although her friend had brief stomach issues, fortunately there appeared to be no permanent damage.

Miller was arrested and charged with possession of known biological toxins without a government license; no charges relating to attempted poisoning were made. At her sentencing the judge rebuked Betty Miller for the "callousness with which she took others' lives into her hands," pointing out that ricin poisoning was "so serious that it's considered a weapon of mass destruction." But the judge also acknowledged the steps that Betty had taken to seek out mental health treatment. In the end Miller paid Wake Robin ninety thousand dollars for hazardous cleanup costs, and the judge imposed a further ten-thousand-dollar fine. She was sentenced to time served with five years of probation, and mandatory mental health treatment. Wake Robin subsequently released a statement to the effect that authorities had declared the campus safe from ricin contamination, and that management would continue to focus on the residents' comfort and privacy. Among the steps taken to protect its residents was the removal of all castor plants from its landscaping.

RICIN AS THERAPY

Just over a century ago, Nobel Prize–winning scientist Paul Ehrlich came up with the concept of the "magic bullet": a drug that would go straight to kill the intended target cells, such as cancer cells or infected cells, without affecting any surrounding normal cells. Certainly ricin is incredibly effective in killing cells, but it lacks the selectivity toward specific cell targets. To overcome this problem, scientists have looked for ways to shuttle ricin to target "bad" cells. The most widely used shuttles are antibodies, which have an incredible ability to seek out and latch on to specific proteins on the surface of cells. In 1976 a group working at the Boston University School of Medicine discovered a way to link the toxic ricin A chain to antibodies, without hampering the effectiveness of either component. In 2016 a clinical study was performed using antibodies that were designed to be attracted to cancerous cells in the bladder. As the antibodies bound specifically to the cancer cells, they took along the ricin A chain. Like a Trojan horse, the cancer let the antibody-ricin complex inside its cells, where the ricin went on to kill the tumor. The realization of Ehrlich's magic bullet now seems a real possibility. Thus cutting-edge therapeutic uses of ricin would seem to rehabilitate the reputation of even the deadliest of substances.

At present there is no cure for ricin poisoning, but recent work in France has identified some promising leads. Searching through a library of 16,480 different chemical compounds, two were found to protect mice from a lethal dose of ricin. More research is needed before these compounds can be made into drugs for humans, but it is possible that these drugs may also prove useful against other virulent toxins, such as Shiga toxin, that enter the cell in a similar way to ricin. While the name Shiga toxin may not be familiar, the bug that makes it and the effects of the toxin are well known. Shiga toxin is made by certain strains of *E. coli,* and causes severe, often bloody, diarrhea, and is the reason for so many *E. coli*–related food recalls.

In the next chapter, we will introduce a drug that was a carefully guarded secret for many years. It is yet another example of Paracelsus's notion that a small amount can be a tonic, but a larger amount becomes a toxin. This journey will take us right to the heart of things, as we examine how the heart beats to move blood around the body.

. { 6 } .

Digoxin and the Angel of Death

A large dose of digitoxin thrown suddenly on the circulation by intravenous injection would cause sudden death by quick palsy of the heart.

—Agatha Christie, *Appointment with Death,* 1938

THE STORY OF DIGOXIN AND DIGITALIS

Foxglove plants can be found throughout Western Europe, Western and Central Asia, Australasia, and Northwest Africa. Though the plants can grow wild, they are often cultivated and grown in gardens, due to their spectacular spikes of flowers in a whole range of colors and hues. The name "foxglove" is something of a mystery, and many theories have been proposed as to its origin, but what is known is that as a plant name it appears in a manuscript written almost a thousand years ago, in 1120.[1] Despite the attractiveness of the plant, it harbors a dark secret. Its flowers may be appealing, but its leaves contain a deadly poison. The poison's reputation has been somewhat rehabilitated over the last two hundred years, when it was discovered that it could be used to treat heart failure. Unfortunately, the same properties that make foxglove medically useful can also be used to kill.

Foxgloves, or more scientifically, plants of the genus *Digitalis*, contain chemicals called glycosides that, like plant alkaloids, are made as a deterrent against the plant being eaten by animals. When extracted from foxglove plants, the glycosides have specific and dramatic effects on the heart, and are therefore known as "cardiac glycosides." Just as atropine got its name from *Atropa belladonna*, so too does the plant *Digitalis* give its name to the poison from its leaves, also known as digitalis. In reality digitalis is a mixture of different glycosides, the most important of which are confusingly called digitoxin and digoxin. Digitoxin is now rarely used medicinally, since not only is it less effective than digoxin, but it has more side effects than digoxin and takes longer to eliminate from the body. (Moreover, patients are likely to take exception to being injected with something that ends in *toxin*.) While digitoxin has no current use, its counterpart, digoxin, is used regularly in today's hospitals.

The modern medical use of digoxin starts with William Withering, an eighteenth-century physician living in Shropshire, England. Withering wanted to find a way to treat his patients who were suffering from dropsy. Dropsy, now known as edema, has several causes, but an important one is a weakened heart, or heart failure. Despite its terrible suggestion, heart failure does not mean the heart has stopped working, but rather that it is working less efficiently. With a weakened or inefficient heart, not only is oxygenated blood transported poorly throughout the body, but the heart muscle thickens and becomes stiffer. Poor blood circulation also causes the kidneys to stop functioning properly. Since one of the roles of the kidneys is to eliminate excess water from the body, when the kidneys are not functioning correctly, fluid accumulates around soft tissues. This often results in swelling of the lower legs, ankles, and feet, leading to aching and tender limbs, a process called edema. Fluid can also accumulate in the space surrounding the lungs, making it harder for them to inflate properly; this leads to shortness of breath and fatigue. Taken together, such symptoms are classically described as congestive heart failure (CHF).

Withering had heard of a woman living out in the woods who concocted herbal remedies for heart problems. Indeed, after taking her herbal treatments, many of Withering's dropsy patients made a remarkable recovery. Intrigued, Withering persuaded the woman to give him some of the remedy, and learned that the main ingredient was foxglove extract. Withering then began a series of experiments using foxglove to treat patients, carefully starting with a low dose and gradually building up till the patient improved. Withering's studies are now regarded as the first clinical trial of a drug, and he was a trailblazer in drug discovery.

Although foxglove in small amounts appeared beneficial, Withering also noted that at higher amounts, toxic effects developed. One of the reasons that even today, patients taking foxglove (or rather its modern derivative, digoxin) are carefully monitored is that the amount of digoxin giving therapeutic benefit, and the amount causing toxic side effects, are very similar.

THE ANGEL OF DEATH

On March 2, 2006, Charles Cullen was brought into the courtroom in Somerset County, New Jersey, where the presiding judge handed down his sentence. Cullen sat immobile, refusing to speak and staring at the ground as his victim's families recounted the devastating toll his actions had caused. Even the exasperated judge could not elicit a response from Cullen, and finally sentenced him to eleven consecutive life terms in the state prison, without the possibility of parole for 397 years.

During questioning by police, Cullen had confessed to killing up to forty patients during a sixteen-year killing spree at seven different hospitals; many investigators believe the real victim count is closer to four hundred. All of Cullen's victims had died during their stays in hospital units where he'd worked as a critical care nurse. The victims had little in common, ranging in age from twenty-one to ninety-one, male and female. Some were in critical condition, and some were about to be discharged and sent home. At his trial the newspapers dubbed him "the Angel of Death," though Cullen was no mercy killer.

Charles Cullen was born in West Orange, New Jersey, in 1960, the youngest of eight children. Life never seemed to be fair to Cullen, from the death of his father when he was an infant to his mother being killed in a car accident when he was seventeen. Indeed, Charlie, as he liked to be called, often lamented that his childhood was miserable. After his mother died, Charlie dropped out of school and enlisted in the U.S. Navy, eventually serving aboard a submarine, where he was a member of the team operating the sub's Poseidon ballistic missiles. During this time, his mental instability started to reveal itself to those around him. On one bizarre occasion Charlie turned up for his shift not in his regular submariner's uniform, but wearing a surgical gown, mask, and gloves he had stolen from a medical cabinet. Appropriately, his commanding officers deemed he was not the kind of person they wanted around ballistic missiles, and Charlie was transferred to a surface supply ship before finally receiving a medical discharge in 1984. Back in civilian society, Charlie enrolled at the Mountainside Hospital School of Nursing in New Jersey.

After graduation Cullen would go on to work at eight different hospitals, and be suspected of harming patients at six of them, but those suspicions never made it to subsequent employers. Many times, rather than involve the police, hospitals instead performed flawed and perfunctory internal investigations that yielded inconclusive results. Hospital administrators were concerned that if it became known they had employed a killer nurse, they would be open to all manner of litigation. So, rather than pursue the matter, Cullen was simply allowed to resign each time suspicions rose around him, making him somebody else's problem. Given the severe nursing shortage in the 1990s, Cullen was easily able to obtain new employment, particularly because he sought out the unpopular graveyard shifts.

In 1993 Cullen was working at Warren Hospital in Phillipsburg, New Jersey. He was employed in the position he would seek out at many other hospitals, one that made it easy to kill without notice: the cardiac and intensive care units, where death was not unusual. Significantly, these units were also where Cullen could get easy access to digoxin.

Mrs. Helen Dean was an elderly patient at Warren Hospital recovering from surgery for breast cancer, and doing well—so well that she was due to be discharged the next day. Helen's son, Larry, had been a faithful visitor during her hospital stay, and seemed never to leave her bedside.

Larry was sitting next to his mother when Cullen entered the room. Larry found this strange; he had been at the hospital every day since his mother was admitted, and pretty much knew every nurse on the floor, but he had never seen Cullen before. However, when Cullen asked him to leave the room, Larry did as he was told, and headed to the cafeteria for a coffee. In Cullen's palm, carefully concealed, was a syringe containing the contents of three ampoules of digoxin; a milligram and a half total, three times the recommended dosage. Helen knew she was getting ready to go home, and was unsure why she was receiving more medicine, but trusting the nursing staff, she allowed Cullen to inject her.

By the time Larry returned, Cullen was gone. "He stuck me!" complained Helen. Pulling up her gown, she showed her son where Cullen had injected her, and there on her inner thigh was a pinprick. Larry called the doctor, who gave the wound a cursory look and suggested it was probably a bug bite.

The next morning, Helen was due to be discharged, but that was not going to happen. She was suddenly very ill, sweating profusely, and exhausted. Helen's heartbeat became very erratic, and when it finally stopped, she could not be resuscitated. Completely distraught, Larry sought out his mother's oncologist, who confirmed that Helen had not been scheduled for any injections. When Larry complained to the other nurses, he learned the name of the mysterious male nurse. Larry Dean was now convinced that something had happened to his mother, and that Charles Cullen was responsible. He called the Warren County prosecutor's office, saying that his mother had been murdered, and that he knew who had done it.

Cullen was questioned by Helen Dean's oncologist, hospital administrators, his nursing supervisors, and investigators from the Warren County prosecutor's major crimes office. All of them wanted Cullen to go carefully through the events leading up to Dean's death. He denied ever having injected her with anything, even passing a polygraph test. Meanwhile the medical examiner's office had taken a sample from around the injection site on Dean's leg, and proceeded to perform a battery of tests. Nearly one hundred potentially lethal chemicals were tested for, but for some reason digoxin was overlooked. No chemicals were found, and the death was ruled as natural. Larry remained convinced that his mother had been murdered, and spent the last seven years of his life trying to prove Cullen's guilt. When Larry Dean died in 2001, still trying to prove that his mother had been murdered, his late mother's blood and tissue samples were found in his home freezer. It would be another five years before Cullen pleaded guilty to Helen Dean's murder.

By December 1998 Cullen was working as a nurse at Easton Hospital in Philadelphia. Seventy-eight-year-old Ottomar

Schramm was a retired Bethlehem Steel worker transferred from his nursing home to Easton Hospital following a seizure. Schramm's daughter, Kristina, thought nothing of the male nurse who came in to check on her father, although she was a little concerned when the nurse said he had to take Schramm out of his room to "perform some tests"; the syringe he carried was "just in case [her father's] heart stopped." Since most patients in a critical care unit are hooked up to IV drips to maintain hydration and make it easier to administer drugs, Cullen merely took advantage of what was already there, injecting the digoxin into the saline bags that slowly and relentlessly dripped the medication into Schramm's bloodstream.

The next time Kristina saw her father he looked very unwell, far worse than he'd looked when he first entered the hospital. His heart rate was erratic, speeding up and slowing down with no clear pattern. His blood pressure plummeted, and he entered an inexorable downward spiral. Kristina received a rather strange phone call from her father's primary care physician, informing her that someone at the hospital had ordered a set of unauthorized blood tests for her father, tests that showed the presence of digoxin. Digoxin had not been prescribed to Ottomar Schramm; in fact, digoxin was off-limits because Schramm had a pacemaker. Not only was digoxin present in Schramm's blood, but the levels were off the charts. At 1:25 a.m. Kristina received another phone call. Her father's blood was still positive for digoxin, and he was now dead.

HACKING THE DRUG MACHINE

Although Cullen occasionally used other medications, like insulin and lidocaine, to kill his patients, his preferred drug became the powerful cardiac medication digoxin. Digoxin was readily available in critical care units, and Cullen had found a way to hide his acquisition of the drug.

Medications were not simply kept in a locked cupboard, but were held inside mobile computerized cabinets called Pyxis MedStations. These were essentially large metal cash registers, with a computer monitor and keyboard affixed to the top; instead of cash, however, the Pyxis dispensed drugs. Hospital administrators liked the Pyxis because the machine efficiently tracked each nurse's drug usage, linked each drug withdrawal to a particular patient account, simplified billing, and notified the pharmacy when a particular drug was running low and needed restocking. As with any device, as soon as it is put in use, someone will try to find ways to work around the system and exploit its weaknesses. Charles Cullen had worked on nuclear submarines, so he was no stranger to technical devices.

Surprisingly, there were no records indicating that Cullen had used Pyxis to obtain digoxin. Cullen had realized that if he placed an order for digoxin for his patients, then quickly canceled the order, the drug drawer nonetheless popped open. He could then simply remove digoxin without any recording of his withdrawal in the system. Later investigations revealed that, although there were no records of Cullen taking digoxin from Pyxis, he had an unusually high number of canceled orders. When Cullen sensed that investigators were closing in on his hacking, the canceled orders suddenly stopped; unfortunately the killings did not. Instead Cullen seemed to be ordering a large amount of Tylenol. But why was he going to all the effort of logging into Pyxis to obtain one Tylenol pill at a time? It was not until another nurse used Pyxis to order acetaminophen that Cullen's scheme became clear. As the nurse pressed Enter, the drug drawer popped open, and there alongside the acetaminophen (the chemical name for the drug in Tylenol) was digoxin. Acetaminophen and digoxin shared the same A-to-D drawer; Cullen ordered one but withdrew the other.

CULLEN IS CAUGHT

In September 2002 Cullen was hired as a critical care nurse at New Jersey's Somerset Hospital. Astonishingly, the Human Resources Department at the hospital was unaware of Cullen's dark past, had no knowledge that he had been fired or encouraged to resign from six other hospitals, or that he had been under investigation for harming patients. His new employers had no idea that Cullen now had dozens of murders under his belt, and would commit a dozen more once he started at Somerset.

The Reverend Florian Gall was admitted to the Somerset critical care unit with swollen lymph glands and a fever over 100° F, both indicative of a massive bacterial infection. Bacterial infiltration of the lungs had led to pneumonia, and Gall was hooked up to a ventilator to help him breathe. His heart was also showing signs of atrial fibrillation—the chambers of the heart were contracting too quickly, before they were fully filled, leading to inefficient movement of blood through the lungs and body. A cardiologist had prescribed digoxin to slow Gall's heart. Of course, that would be the effect of the *right* dose of digoxin.

At first Gall appeared to improve on the medication, but by the middle of the night, he was struggling to breathe. In contrast to its normal steady rhythm, Gall's heartbeat became irregular, quivering and palpitating inefficiently. The chaotic contractions of the heart failed to send oxygen around the body, causing a sense of breathlessness. At 9:32 a.m. on the morning of June 28, the reverend's heart suddenly and unexpectedly stopped altogether. The crash team immediately swung into action, and for the next half hour worked to revive Gall. Despite the resuscitation team's best efforts, Gall's heart was unresponsive, and at 10:10 a.m. he was pronounced dead.

Blood work taken at the time of Gall's death showed that his digoxin levels were exceptionally high, and his lab reports told a disturbing story. On June 20 Gall's digoxin levels were 1.2, on the twenty-second, 1.08, and 1.33 on the twenty-third; blood drawn at dawn on the twenty-eighth showed Gall's digoxin level had leaped to 9.61. Anything over 2.5 is considered toxic.

Dr. Steven Marcus was the director of New Jersey's Poison Control Center. He became aware of the digoxin overdose when a pharmacist from Somerset called him with questions about how the digoxin levels could have gotten so high so quickly. The blood digoxin levels were so unusual that Marcus immediately suspected foul play. He set up a phone call with Somerset Hospital administrators, arguing that they needed to call the police. The hospital administration was worried that contacting the police would "throw the whole institution into chaos,"[2] and wanted to carry out an internal investigation before rushing to any kind of judgment. Somerset Hospital did finally agree to contact the police, but only after a three-month delay.

Somerset County detectives met with hospital managers to discuss the deaths that had occurred in the critical care unit. When the detectives asked to see the hospital's Pyxis records for the critical care unit, the hospital administrators said it was pointless, as records were stored for only thirty days. In fact Cardinal Health, the manufacturers of the Pyxis machine, informed detectives that, contrary to what the hospital said, records were not deleted after thirty days. Now, with crucial evidence of Cullen's unusual Pyxis requisitions, detectives informed Somerset Hospital that Cullen was their main suspect; the hospital's response was to distance itself by firing Cullen.

With Cullen fired, and the hospital less than cooperative, detectives approached a nurse in Somerset's critical care unit who was a friend of Cullen's and had often worked the night shift with him. She proved invaluable in collecting enough evidence to allow the police finally to arrest Cullen on December 12. During interrogation Cullen confessed to his long killing spree, which eventually led to his 2006 sentencing.

Every time a hospital had become suspicious of deaths surrounding Cullen, their main concern was to get rid of him. Each time Cullen resigned, he was given a neutral reference that enabled him to get a nursing position at another hospital, allowing the pattern of deaths, suspicion, resignations, and reference letters to repeat itself over and over. It will never be known how many lives could have been saved had administrators been more concerned for patient safety than potential lawsuits. In 2005 the New Jersey governor signed into law the State Health Care Professional Responsibility and Reporting Act, which requires hospitals to notify regulatory bodies of any suspicious activity by health care employees, and to undertake criminal background checks of all health care licensees. The new regulations are also referred to as "the Cullen Law."

PROBLEMS OF A BROKEN HEART

The heart is a remarkable organ, contracting roughly 4,800 times each hour to move blood around the body. Over the course of a year the heart beats 42 million times; if you live to be eighty years old, that adds up to more than three billion heartbeats. Each day approximately two thousand gallons of blood move through the heart, in contrast to the typical motorist whose car goes through only six hundred gallons of gasoline in a whole year.

We tend to think of the heart as a simple pump, shunting blood around the body. In fact the heart actually comprises two pumps. The right side of the heart takes deoxygenated blood from around the body and pushes it through the lungs, where the red blood cells pick up their load of oxygen. Blood from the lungs then returns to the left side of the heart so it can be pumped to the rest of the body carrying its oxygenated red blood cells. Each side of the heart is further divided into smaller atria (Latin for "entry rooms"), and larger ventricles (Latin for "chambers"). Blood first enters the atria before being squeezed into the ventricles, then pumped either to the lungs (right side) or the body (left side).

To work efficiently the contraction of the atria has to be perfectly timed with that of the ventricles, so that the atria are finished filling the ventricles with blood before they in turn contract, the system resets, and a new round of contractions is initiated. Electrical signals through the heart are responsible for these coordinated contractions, and it can be imagined how easily the atrial and ventricular contractions can become disorganized and chaotic when the electrical signaling breaks down. Remarkably, each round of coordinated contractions takes less than a second. The contractions of the ventricles are so strong that if the aorta (the main artery leaving the left side of the heart) is severed, say by a stabbing, blood can be squirted up to ten feet from the wound.

HOW DIGOXIN TREATS AND KILLS

Digoxin works in two ways to treat heart failure; it intensifies each round of contractions, and slows down the electrical signaling in the heart. Most of the heart is made from specialized cells called cardiac myocytes, and these are the cells that do the actual contracting to squeeze out blood from the heart. One thing all muscle cells, including the heart, require in order to work properly is calcium. We tend to think of calcium as important only in teeth and bones, but it has lots of functions in the body, including muscle contraction.

The importance of calcium in heartbeats was first discovered by Sydney Ringer in the 1880s. Ringer was looking for ways to keep hearts beating longer after they were removed from frogs so they could be used for closer study. What Ringer found was that, unless calcium was present in the bathing solution, the hearts failed to beat properly. With calcium, they would beat contentedly for up to five hours. Ringer's early experiments on keeping hearts working would be vital to the studies that Otto von Loewi would perform a few years later to end the "soups and sparks" debate (see chapter 2).

One of the roles of cardiac glycosides, like digoxin, is to increase the amount of calcium inside heart muscles to support contraction. The more calcium present, the stronger the contraction. Digoxin accomplishes this in a somewhat indirect path by interfering with the actions of two proteins embedded in the cell membrane. One is the sodium pump, and the other is the sodium-calcium exchanger.

The sodium-calcium exchanger works just as its name suggests. As sodium comes into the cell, it is obliged to kick out calcium in exchange. The more sodium comes into the cell, the less calcium is left inside the muscle cell to support contraction. But what if there were a way to decrease the amount of sodium coming into the cell?

Enter digoxin. Digoxin stops a protein called the sodium pump from working. One of the jobs of the sodium pump is to supply the sodium-calcium exchanger with sodium. No sodium pump means no sodium-calcium exchanger. More calcium in the heart muscle means that the heart beats stronger and more efficiently, which is extremely effective for patients suffering from heart failure. However, as we will see, digoxin can have adverse side effects because the sodium pump is found in nearly all cells of the body. Of course, the trick is to slow the sodium pump down just enough, without completely stopping it altogether. Because of this the symptoms of digoxin poisoning can be varied, including dizziness, confusion, nausea, vomiting, and, as we shall see, blurred vision.

The effects of digoxin on calcium levels are not the only way digoxin helps a heart. As we have learned, the heart has a system of electrical connections in place to make sure the atria and ventricles contract in a coordinated fashion. In some cases the electrical signaling can become disorganized, and inappropriate messages cause the heart to beat erratically in an inefficient manner. Atrial fibrillation is quite a common condition, and occurs when there is a rapid and irregular contraction of the atria. It can also result in the atria and ventricles contracting independently.

Digoxin slows the electrical signaling across the heart, essentially calming it down and helping restore the coordination of contractions. By increasing the level of calcium in the heart muscle cells, and by calming the heart's electrical signaling, cardiac glycosides like digoxin can strengthen the heart's contractions. The stronger the contractions, the more efficient each beat becomes, and the symptoms of congestive heart failure slowly improve.

But digoxin has what is referred to as a narrow therapeutic

window. The difference between the dose of digoxin that has benefit for the patient and the dose that can cause serious problems is very small. The correct dose of digoxin increases heart calcium by just the right amount to increase muscle contractions, but too much digoxin and calcium levels start to become very high. At these abnormal levels calcium can start causing problems with the electrical signaling in the heart. Digoxin can markedly enhance the signals telling the heart to speed up. As the heart rate becomes faster and faster, it becomes more uncoordinated, leading to erratic contractions and ultimately stoppage. As digoxin is altering the signals going into the heart, it is also affecting signals within the heart. We have seen how the atria must contract before the ventricles, and this is brought about, in part, by a specialized tissue in the heart called the AV node that acts as a relay station for signals from the atria to the ventricles. Toxic levels of digoxin obliterate the function of the AV node; doctors refer to this as AV block. Someone undergoing AV block will experience symptoms of dizziness, shortness of breath, chest pain, and the unnerving sensation of the heart skipping beats. Untreated, the heart will skip beats permanently, as it stops contracting altogether, a condition called cardiac arrest.

As the heart becomes weakened and paralyzed, oxygenated blood is no longer transported throughout the body, and a feeling of shortness of breath follows. The brain experiences blackouts and finally loses consciousness altogether. With the brain dead, and the heart stopped, the digoxin-poisoned victim is pronounced dead.

DIGOXIN AND THE HUNDRED-MILLION-DOLLAR PAINTING

We have seen that the sodium pump is present in every cell of the body. However, this is particularly true in the cells of the retina of the eye, where up to thirty million molecules of the pump may reside in each retinal cell.

There are two types of cells in the retina: the rods and the cones. The rods are responsible for vision in low light, being incredibly sensitive to the lowest of light levels, detecting even a single photon. The tradeoff is that the rods cannot detect different wavelengths of light, and can see the world only in shades of gray. The cones are much less sensitive to light, but come with the enormous benefit of being able to see color. Each cone cell can detect red, green, or blue light, and depending on how much of each cone is stimulated at a time, we can see all the different colors. The brain is amazingly able to integrate all the signals from the rods and cones into a picture of the colorful world around us.

Cone cells are fifty times more sensitive to digoxin than are rod cells, and so digoxin affects the ability to see colors much more than it impacts night vision. One of the most complained-about side effects reported by patients taking digoxin for CHF are visual disturbances, including blurred vision, flickering spots, and xanthopsia (where yellow-green halos are seen around objects).

A striking feature of Vincent van Gogh's famous painting *Starry Night*, worth one hundred million dollars, is the yellow corona surrounding each star. The strong use of yellow characterizes many of the paintings of this creative Dutch artist, including *The Night Café* and *The Yellow House*. Did van Gogh just like the color yellow, or was he influenced by some underlying medical condition?

Among the answers to this question is the suggestion that van Gogh was suffering from digitalis toxicity. It is well known that van Gogh suffered from depression and epilepsy, and at the time there was a general assumption among physicians that if a drug worked well against one kind of condition, it would likely work on others too. Although there is no written evidence that digitalis was ever prescribed for van Gogh, two portraits of his personal physician, including *Portrait of Doctor Gachet*, show the good doctor holding a foxglove plant.

AN ANTIDOTE TO DIGOXIN POISONING

In instances of accidental digoxin overdose or deliberate poisoning, a surprising drug we have already come across can be used. That drug is atropine. Recall that digoxin overdose leads to cardiac arrest, whereas atropine causes the heart to speed up, counteracting digoxin's toxic effects. Another drug more usually employed for digoxin overdose today is an antibody isolated from sheep that seeks out and inactivates any digoxin in the blood. Phenytoin is a widely used antiepileptic drug, yet surprisingly also finds use as a treatment for digoxin overdose by increasing the metabolism of digoxin in the body.

In this chapter we have seen how digoxin knocks out a cell's sodium pump, affecting the levels of sodium and calcium in the cell. Although usually referred to as the sodium pump, it is also known as the sodium-potassium-ATPase, indicating that not only does it affect sodium levels but it also changes potassium levels inside cells. When digoxin inhibits the sodium pump, or sodium-potassium-ATPase, potassium starts to leak out of cells, increasing the amount of potassium in the blood. As we will see in the next chapter, such increases in blood potassium can have their own deadly consequences.

. { 7 } .

Cyanide and the Professor from Pittsburgh

No, no, the drink, the drink! . . . I've been poisoned!
—William Shakespeare, *Hamlet*

A MOST FAMOUS POISON

Cyanide is one of the most notorious poisons around, having a reputation in spy novels and murder mysteries for causing almost instantaneous death. The Queen of Mystery, Agatha Christie, was very much aware of the effects of cyanide, and used the poison to kill off eighteen characters, even naming one of her seventy-five novels *Sparkling Cyanide*. Gumshoe novelist Raymond Chandler, in his most famous book, *The Big Sleep*, employed cyanide-laced whiskey to bump off an informant. Nevil Shute's novel *On the Beach* recounts life in Australia following a devastating nuclear war. In the story, the Australian government hands out cyanide capsules to the population, so that they can quickly and easily kill themselves rather than face a slow agonizing death from the radioactive dust cloud that approaches over

Australia. Similarly, secret agents in spy novels were frequently given cyanide to take in case they were captured. Even Ian Fleming's James Bond, along with other agents, was issued cyanide capsules; although Bond, as would be expected, threw his away.

Real-life uses of cyanide to commit murder or suicide are equally fascinating and appalling. As a murder weapon, cyanide is associated with some of the worst crimes in history. During World War II, hydrogen cyanide was used to gas thousands of prisoners in the death camps of Auschwitz-Birkenau and Majdanek as part of the so-called Final Solution.

When it became apparent that Germany was going to lose the war, a glass capsule containing potassium cyanide was the preferred method of suicide by the Nazi hierarchy, including Heinrich Himmler, head of the dreaded Schutzstaffel (SS), and Hermann Göring, head of the German Luftwaffe (air force). After witnessing his wife, Eva Braun, commit suicide with cyanide, Adolf Hitler swallowed cyanide and shot himself, finally ending the dream of a Third Reich.

More recently, in early 1970s San Francisco, charismatic cult leader Jim Jones garnered a large following. He set up a temple in Redwood Valley, California, where he began preaching that he was the reincarnation of Gandhi, Jesus, Buddha, and Lenin. By the mid 1970s Jones had convinced hundreds of people, including entire families, to move to a new utopia in Guyana at the People's Temple in the eponymous Jonestown. In 1978 concerns began to surface about human rights abuses and harsh punishments surrounding the Jonestown temple. That November, Congressman Leo Ryan, along with other U.S. officials and journalists, traveled to Guyana to investigate the allegations.

At first Jones welcomed the delegation to his compound, hosting a reception for them in the central pavilion of Jonestown, but Ryan was suddenly attacked by a knife-wielding temple member, sustaining several injuries. Despite being wounded he was able to escape along with the members of his delegation, who boarded two planes on an airstrip close to Jonestown. Within a few seconds of boarding the flight, however, gunmen arrived, killing Ryan and four others. Later that day Jones assembled 913 inhabitants of Jonestown, including 304 children, commanding them to commit what he termed a "revolutionary act." Cups of grape-flavored Kool-Aid containing cyanide were handed out to drink. Children were given the drink by their parents, and nurses used syringes to drip the lethal mixture into babies' mouths. Altogether 909 people were killed, a third of them children. The phrase "drinking the Kool-Aid" is an idiom still commonly used in the United States to refer to an individual, or a group, that demonstrates unquestioning obedience or loyalty to an idea or person.

DYEING FOR CYANIDE

The name *cyanide* is derived from the Greek word *kyanos*, meaning "dark blue," by a somewhat circuitous pathway. During the Renaissance, blue pigments were all derived from the semiprecious mineral lapis lazuli. The resulting pigment was incredibly expensive, worth more than five times its weight in gold, and so blue was used very judiciously in artworks.

The solution to the "blue problem" was found by accident in 1704 by color maker and artist Heinrich Diesbach, working in the Germanic Kingdom of Prussia. Diesbach was in a rush to manufacture a batch of Florentine lake, a red pigment made from boiled cochineal beetles, alum, iron sulfate, and potash. He had everything he needed except for the last ingredient, and, being a bit short on cash, decided to buy cheap materials. More than willing to help in this scam was alchemist Johann Konrad Dippel,[1] who was familiar with the concept that would be espoused 150 years later by P. T. Barnum: "There's a sucker born every minute!" Dippel did in fact have potash in his inventory, but it had been contaminated with animal oil (a disgusting amalgam of blood and various other animal parts) and was due to be thrown out. Sensing an opportunity to avoid financial loss, Dippel sold the tainted potash to Diesbach, each believing he had gotten the better end of the deal.

Back home, Diesbach proceeded to stew his pigment mixture with iron sulfate and the cheap contaminated potash, but instead of the bright red color he was expecting, all he got was a very dismal muddy rust color. He theorized that if he heated the dye longer to concentrate the color, he would end up with the desired red. In fact, what Diesbach obtained was first a purplish color, and then a deep blue. Diesbach rushed to Dippel to find out what he had really been sold.

Together Diesbach and Dippel realized the commercial opportunities in this new synthetic dye, and immediately joined forces to make batches of blue dye to sell to artists in the Prussian court. Diesbach called the new dye Berlin blue, but English chemists would change the name to the more familiar Prussian blue, since the new blue color was by then available in sufficient quantities to dye the uniforms of the Prussian army.

Later chemical analysis would reveal that right in the center of each molecule of blue pigment was cyanide. So why didn't the Prussian army all die from cyanide poisoning? Certainly cyanide on its own is very dangerous, but when cyanide becomes incorporated into larger molecules it loses its lethality, and the blue pigment was one such safe molecule.

The new blue dye was an immediate sensation, with painters wanting to get in on using the new blue colors in their artwork. The Venetian artist Canaletto was an early adopter of the novel pigment, using it for his dramatic skies in *Westminster Bridge*, painted in 1747. Even two hundred years after its creation, Pablo Picasso could not have had his Blue Period without it, nor could van Gogh have created his *Starry Night* (also featured in the chapter on digoxin). Ironically, even though *Starry Night* is now priceless, van Gogh could not have afforded to even paint the picture with so much blue had it not been for Diesbach and Dippel.

Eighty years after the discovery of Prussian blue, French chemist Pierre-Joseph Macquer and Swedish chemist Carl Wilhelm Scheele, apparently bored one afternoon, decided that it would be amusing to mix Prussian blue with acid and heat it, just to see if anything happened. What they obtained was iron oxide, better known as rust, and a curious colorless vapor that was almost undetectable, save for the faint smell of almonds. The gas they'd produced was hydrogen cyanide, and when cooled and dissolved in water, it yielded a very strong acid. The acid was known as Prussic acid, although later chemists preferred to call it by its more proper chemical name: hydrocyanic acid.

Cyanide is a simple molecule, consisting of only two atoms, one of carbon and one of nitrogen, bonded together. The cyanide molecule binds to many metals, including iron, cobalt, and gold. In fact cyanide is one of the few chemicals that will react with gold, and is the reason why cyanide is used in the extraction of gold from its mineral ores. Poisonous cyanide can exist in solid, liquid, and gas forms. Solid cyanide is found as a white crystal, often paired with sodium or potassium, to make sodium cyanide and potassium cyanide. Cyanide can also pair with hydrogen to make hydrogen cyanide. When cooled, hydrogen cyanide is a pale blue liquid, but it is very volatile, and even at room temperature exists mostly as a gas, with a faint smell of bitter almonds. All these forms of cyanide are deadly: just 50–100 mg (as little as 1/100th of a teaspoon) of potassium cyanide can kill an adult.[2]

Remarkably, when cyanide is not on its own but part of a larger molecule, some forms are totally innocuous. When found in Prussian blue dye, for example, the cyanide is quite safe, allowing Gainsborough to use the pigment in his *Blue Boy* without immediately keeling over dead. The safety of cyanide when it is present in certain large molecules also means that anyone taking a daily multivitamin is likely swallowing cyanide tightly (and safely) bound to vitamin B12 (cyanocobalamin). Medications for depression and acid reflux, taken by millions of people around the world, both contain safely bound cyanide.

CYANIDE IN THE DIET

Despite its apparent lethal characteristics, cyanide is found in a surprisingly large number of foods, including almonds, lima beans, soy, spinach, and bamboo shoots. The seeds or pits from plants of the genus *Prunus*, which includes peaches, cherries, apples, and bitter almonds, all contain cyanide. Eating small amounts of cyanide poses no health risk; indeed, most of us have swallowed the occasional apple seed with no ill effect. This is because humans have a mechanism for dealing with small amounts of cyanide in our diet. Almost every cell in the body contains the enzyme rhodonase, which rapidly detoxifies cyanide by converting it into thiocyanate, an innocuous chemical that can be safely filtered by the kidneys and released into the urine. Humans can process about 1 g of cyanide every twenty-four hours. Problems occur only when we overload the body with a sudden large influx of cyanide—especially when the cyanide is intended to kill.

Most murderers will give their victims either sodium or potassium cyanide crystals. Although both are quite soluble, potassium cyanide is ten times more soluble than sodium cyanide. Even so, a small amount of either dissolved in a cup of coffee or a glass of wine is more than enough to kill. The small amount required means it has no odor or taste to alert the victim. When the cyanide crystals are swallowed, they come into contact with stomach acid. Here the sodium or potassium cyanide becomes Prussic acid, causing severe chemical burns. The presence of caustic burns to the stomach, but not the esophagus, imply that the victim did not drink anything caustic, but that the cause was generated in the stomach—a key indicator of cyanide. When solid or dissolved cyanide crystals encounter stomach acid, hydrogen cyanide gas is also formed. The gas can be absorbed and enter the bloodstream to be transported throughout the body. In essence, a victim is eventually killed by solid, liquid, and gaseous cyanide.

HANGED BY TELEGRAPH CABLE

It has been argued that the job of a defense lawyer is not necessarily to prove a client innocent, but to sow seeds of doubt in the minds of the jury. In addition to seeds of doubt, John Tawell's defense would also introduce seeds of apples.

John Tawell worked in London's early Victorian Quaker community, supporting a wife, Mary, and two young children. Infant mortality rates were astonishingly high, and both of Tawell's sons died in childhood. Heartbroken, Mary soon became ill herself, likely from a combination of grief and polluted air. Tawell employed an attractive young nurse named Sarah to look after his wife, seeing to her daily care. Despite the nursing care, Mary, too, was soon dead, just a few months after her youngest son. Barely was Mary in her grave when Tawell started an affair with his wife's nurse, producing two illegitimate children.

Tawell set up his mistress, and their children, in a house in Salt Hill, near Slough (a large town some twenty miles west of London), making regular visits and paying her £1 a week (roughly £80 or $100 today) for housekeeping. But by 1843 Tawell was experiencing severe financial difficulties, and even though upkeep for his mistress was a mere £1 a week, he took Benjamin Franklin's maxim "A penny saved is a penny earned" to heart. Without Sarah, he could save 240 pennies a week.

On New Year's Day, 1845, Tawell went into a pharmacy and bought two bottles of Steele's acid, a treatment for varicose veins that just happened to be made from Prussic acid. Tawell then went to Paddington Station, where he boarded a train to Slough to meet with his mistress, taking with him a bottle of stout he had bought at a local inn. What happened in the next hour or so is still a mystery, but Tawell likely distracted Sarah long enough to pour Prussic acid into her beer. A short while later loud shouting, groaning, and moaning were heard by Mrs. Ashley, the next-door neighbor, who looked out her front window to see Tawell, whom she knew was a frequent visitor to Sarah's house, walking swiftly toward the train station.

Concerned for Sarah's well-being, the neighbor rushed next door to find Sarah writhing in agony on the floor and frothing at the mouth. Mrs. Ashley quickly summoned a doctor, but it was too late: Sarah was dead before he arrived. Unfortunately for Tawell, the doctor was not the only one who responded to

Mrs. Ashley's cries for help. The Reverend E. T. Champnes ran to Slough station with a description of Tawell, trying to stop him before he could escape. Alas, the reverend arrived at the station just in time to observe Tawell boarding the 7:42 p.m. train to London's Paddington Station, a ride of about an hour.

It was impossible to catch up with Tawell now, and once in London he could disappear into the crowds. What Tawell didn't know was that Slough was one of a few train stations equipped with the new electric telegraph system. In a flash the reverend realized that a telegraph communication would arrive at Paddington long before Tawell did. The message he sent read: "A murder has just been committed at Salt Hill and the suspected murderer was seen to take a first class ticket to London that left Slough at 7:42 p.m., he is in the garb of a Kwaker. He is in the last compartment of the second first class carriage." (The telegraph system did not have the letter Q, and so *Quaker* was spelled phonetically.)

At Paddington the message was relayed to the on-duty sergeant, who put on a long plain overcoat to conceal his police uniform and calmly waited for Tawell's train to arrive. Keeping him under observation, the sergeant followed Tawell to his house. Satisfied that Tawell was staying put, the sergeant headed off to make his report to Inspector Wiggins of the Metropolitan Police.

Tawell was arrested the next day and sent to trial for the capital murder of Sarah Hart. Due to the use of the telegraph in capturing Tawell, his trial garnered large national interest. Tawell's defense lawyer was Sir Fitzroy Kelley, a lawyer well versed in commercial law but with little knowledge of criminal jurisprudence.

The main thrust of Kelley's defense argument was that Sarah had indeed died from cyanide poisoning, but that was due to her consumption of a large quantity of apple cores, and not from any intervention by Tawell. When the prosecution argued that Sarah would have needed to eat several thousand apple seeds to ingest a lethal quantity of cyanide, the defense crumbled. The trial lasted two days, but the jury took only thirty minutes to consider their verdict before finding Tawell guilty of murder. Tawell was executed by public hanging outside the court, with more than ten thousand witnesses on hand. Londoners referred to the telegraph wires as "them cords that hungth John Tawell." And what of Tawell's defense lawyer? He earned the nickname "Apple-pips Kelley."

HOW CYANIDE KILLS

Whether a murder victim inhales cyanide gas or swallows sodium or potassium cyanide dissolved in a drink, cyanide kills in exactly the same way. Once in the body, cyanide can stick to hemoglobin in red blood cells and hitch a ride, to be rapidly distributed by the blood. However, cyanide binds quite poorly to hemoglobin, and causes its devastating effects not by affecting the blood but by hopping off hemoglobin and entering into cells. Once inside the cells of the body, cyanide disrupts their ability to generate the energy needed to live.

Located deep within each of our cells are mitochondria, small rod-shaped structures that act as tiny power plants to generate the chemical energy adenosine triphosphate (ATP), which keeps us alive. Typically there are between one hundred and two hundred mitochondria present in each cell, depending on how much energy the cell requires. Liver cells, for example, require a lot of energy and can contain upward of two thousand mitochondria per cell. Red blood cells, which are mostly bags of hemoglobin, have very low energy requirements, and lack any mitochondria. Despite its importance in fueling every aspect of the body, only limited amounts of ATP are ever stored.

Essentially mitochondria perform the opposite function of leaves in trees. In plants, leaves use the energy in sunlight to combine water and carbon dioxide to produce glucose. In animal cells, mitochondria break down glucose from our food by reacting it with the oxygen we inhale, to make carbon dioxide and water and release energy; this time in the form of ATP. In this circuitous way, humans and all animals are essentially able to harness energy from the sun.[3]

Buried in the membrane lining of the mitochondria are a series of linked proteins that make up the so-called electron transport chain. It is here that the oxygen we breathe is actually used in the process of making ATP. One component of that chain is a protein called cytochrome C. Sequestered at the heart of cytochrome C, and vital to its function, is an atom of iron.

Cyanide's lethality is due to its ability to bind tightly to the iron atom at the center of cytochrome C, rendering the protein dead. Once inactivated, cytochrome C cannot utilize oxygen in the final step of the chain, and the whole production of ATP grinds to a halt.

Because they rely heavily on a continuous supply of ATP, cells of the central nervous system and the heart are immediately affected in the case of cyanide poisoning. As the central nervous system shuts down, the victim develops headaches and nausea before becoming unconscious, and slowly entering a deep coma. Further loss of ATP energy in the brain occurs until levels are finally depleted, and brain death is inevitable. As ATP levels in the heart drop, the heart slows down and becomes erratic. The pulse becomes so weak that it cannot be found, and eventually the heart stops altogether.

Despite its similar-sounding name, cyanosis is not associated with cyanide poisoning. Cyanosis refers to the blue color associated with deoxygenated blood, which is why blood in the veins appears blue. In contrast, since cyanide-bound cytochrome C can no longer use oxygen, the hemoglobin in the blood remains oxygenated.[4] Therefore, one of the symptoms of cyanide poisoning is bright red oxygenated blood, which gives the skin a ruddy appearance.

DEATH ON THE ALLEGHENY

Nestled between the Allegheny and Monongahela Rivers before they join to form the Ohio is the University of Pittsburgh Medical Center (UPMC), a world-renowned hospital and top-tier medical research facility. In May 2011, two well-known Boston neuroscientists, Dr. Robert "Bob" Ferrante and his wife, Dr. Autumn Cline, settled into their new jobs on the faculty of the University of Pittsburgh School of Medicine. Attracting them from

Massachusetts to Pittsburgh was certainly a coup for the university's administration, since Ferrante was bringing with him several million dollars in research grants. Ferrante was a professor of Neurological Surgery, whose research focused on neurodegenerative diseases such as amyotrophic lateral sclerosis (ALS), better known as Lou Gehrig's disease. Less than six months after joining the faculty at Pittsburgh, Ferrante was chosen as the first recipient of the Leonard Gerson Distinguished Scholar Award for his research.

Dr. Cline, too, settled into her new life as chief of Women's Neurology. She was a well-loved physician, board certified in clinical neurology, and specialized in seizure disorders during pregnancy. The move to Pittsburgh, along with the promotion and the ability to run her own program, gave Autumn the ability to work her own hours, and the short fifteen-minute walk from their house meant that she could spend more time with her six-year-old daughter.

But Wednesday April 17, 2013, would be a long day. The time was just after 11:15 p.m., and Autumn was exhausted from a grueling fifteen-hour shift. As she got ready for the half-mile walk home, she messaged her husband that she was on her way. Thirty minutes later Robert Ferrante was calling 911.[5]

> **911:** Allegheny County 911. What's the address of your emergency?
>
> **Ferrante:** Hello. Please, please, please. I'm at 219 Lytton Avenue. I think my wife is having a stroke.

Ferrante explained that Autumn had just arrived home and had collapsed in the kitchen. As the 911 operator tried to get more information, Autumn's groans could be heard in the background. Curiously, despite their house being only a few hundred yards from the main medical school hospital where Ferrante and Autumn worked, Ferrante insisted that the paramedics take his wife to Shadyside Hospital, a mile and a half away

Twelve minutes later the paramedics arrived and rushed into the kitchen, where Autumn lay unresponsive on the floor.

"She came in complaining of a headache, then just collapsed," said her husband.

A preliminary evaluation showed that Autumn was still breathing and had a pulse. The paramedics asked about a plastic bag of white powder sitting on the counter, wondering if it had anything to do with Autumn's condition. Ferrante replied that it was creatine his wife was taking for infertility.

Suddenly Autumn's condition worsened, her pulse and blood pressure dropping fast. She was loaded into an ambulance. The paramedics ignored Ferrante's request to be taken to Shadyside Hospital, and was instead driven to the much closer emergency entrance of UPMC Presbyterian Hospital, just over an hour after she'd walked out of that same building. Was Ferrante hoping that the chances of his wife's recovery would be lessened by driving to a hospital farther away?

Autumn lay in an emergency room, clearly struggling to breathe. Her blood pressure had continued to drop and was hovering at 48/36. To keep her breathing, she was intubated and hooked up to a respirator. Her symptoms suggested that she was having a brain hemorrhage, but a CT scan failed to show any abnormalities. And although her heart rate was incredibly low, there was no evidence of any change in the electrical activity of the heart. Doctors injected her with adrenaline, just to keep her heart going.

The ER team had no clue what was wrong with Autumn. A central line was inserted into Autumn's jugular vein, making it easier to administer drugs and to withdraw blood for testing. Curiously, instead of the dark red color that deoxygenated blood from veins should be, Autumn's venous blood was bright red, the color of oxygenated arterial blood. In fact the level of oxygen in her veins was more than double what it should have been; Autumn's cells were unable to use the oxygen being delivered to them.

Despite heroic attempts to save Autumn's life, by 12:31 p.m. on Saturday, April 20, she was dead. The death of an otherwise healthy forty-one-year-old is not normal, and Ferrante was asked to give his permission for an autopsy on Autumn to see if they could determine a cause of death. However, Ferrante was adamant that no autopsy be performed. He was so insistent that several physicians noted his refusal on Autumn's charts.

But Pennsylvania law required an autopsy, whether Ferrante agreed or not. The autopsy, along with blood work taken during Autumn's stay at the hospital, found something surprising and shocking in her blood: cyanide. Not just a small amount of cyanide, but extremely high levels that likely would have knocked her to the floor in seconds. But where had the cyanide come from? There were only three possible explanations: accidental exposure, suicide, or murder.

It's hard to be exposed to lethal cyanide levels accidentally, and suicide seemed unlikely. All of Autumn's colleagues described her as a caring mom and a passionate researcher who was excited about upcoming research projects. Notably, none of the projects she was working on involved any cyanide.

Researchers at most universities typically purchase chemicals and equipment through the university's purchasing department, with a standard turnaround of four to seven days. Bob Ferrante used a different method called a P-card, or purchasing card; essentially a university credit card that allows researchers to order chemicals directly over the phone, usually with a twenty-four-hour turnaround. According to colleagues, the only time Dr. Ferrante ever used his P-card was on April 15, two days before Autumn's collapse. What had Ferrante bought? The only purchase ever made on the credit card by a highly educated brain researcher, and signed with his own signature, was cyanide.

A hunt through Ferrante's internet history, revealing searches for "divorce in Pittsburgh PA" and "detecting cyanide poisoning," helped convince detectives to arrest Ferrante and charge him with the murder of his wife. During the eleven-day trial, the prosecution noted that more than eight grams of cyanide was missing from the newly purchased bottle in Ferrante's lab. Ferrante countered that he had been planning to use cyanide to kill nerve cells in experiments. Cyanide would certainly kill cells in a laboratory, but the approach has all the subtlety of a sledgehammer, with no selectivity for one cell type over another, making Ferrante's assertion somewhat perplexing.

In closing arguments the prosecutor told the jury that Ferrante was a master manipulator, and that if they put the pieces of the puzzle together, they would realize that he had killed his wife because he thought she was going to leave him. On that fateful night Ferrante gave Autumn a poisoned drink, called 911, and stood over her to watch her suffer. After fifteen and a half hours of deliberation over two days, he was found guilty of murdering his wife by cyanide poisoning. Ferrante is serving a life sentence with no possibility of parole.

A TREATMENT FOR CYANIDE POISONING

Despite the lethality of cyanide poisoning, there are nonetheless remarkably effective antidotes. The trick is to get them to the victims in time. Unfortunately cyanide acts so quickly that up to 95 percent of accidental cyanide exposures are fatal. Mouth-to-mouth resuscitation of victims is not an advisable practice, as the rescuer is as likely to inhale hydrogen cyanide gas from the victim's lungs and stomach as to breathe air into the victim. Today, workers whose job requires them to work around cyanide always carry an antidote kit in case of emergency.

One kind of cyanide antidote works by luring away the cyanide from the mitochondrial cytochrome with another, more tempting molecule. Remarkably, one such molecule is taken by millions of people every day in the form of vitamin supplements:

specifically, vitamin B12, or cobalamin. At the heart of vitamin B12 is an atom of the metal cobalt, and cobalt is much more attractive to cyanide than the iron found in cytochromes. In fact, cobalt latches onto cyanide so tightly that if B12 is injected into a cyanide poisoning victim, the cobalt will mop up all the cyanide, rendering it completely inert.

CYANIDE AND THE ARSONIST

It is rare that someone dying of cyanide poisoning is filmed as it happens, yet that is exactly what happened in a Phoenix, Arizona, courthouse in 2012.

Michael Marin was a graduate of Yale Law School, and had enjoyed a lucrative career on Wall Street. He was a thrill seeker who flew his own airplanes, and had even climbed Mount Everest. Marin owned a large estate in Phoenix, with a monthly mortgage payment of $17,250. But by 2012 Marin had long since left Wall Street and was quickly running out of money. That's when, prosecutors claimed, Marin decided to burn down his house and collect the insurance money.

In July 2012 a jury found Marin guilty of arson. As the verdict was read, Marin was facing between seven and twenty-one years in prison. On court video Marin is seen reaching down to his bag and pulling something out, which he brings to his face and appears to swallow. Eight minutes later Marin is seen falling out of his chair, collapsing and convulsing. Investigators believe that Marin had fashioned some sodium cyanide powder he had purchased a year before into a capsule to be consumed in court.

Anyone drinking a glass of wine that has a faint odor of almonds may want to consider who poured their glass. The aroma associated with cyanide is one of its most famous attributes, but it appears not everyone is able to smell cyanide. In one series of experiments, 244 individuals, including parents and their children, were exposed to cotton soaked in either distilled water or a solution of potassium cyanide, and asked if they could smell anything. The initial report does not say whether the subjects were aware, and if aware, willing, that cyanide was involved in the experiment. What is sure is that such an experiment would likely not be allowed in today's health and safety environment. Nevertheless, the results do reveal some interesting things. Between 20 and 40 percent of people are unable to detect the odor of cyanide, with a greater percentage of males than females unable to do so. The ability to smell cyanide also runs in families. Whether this trait has been used to hide cyanide from an unfortunate family member is unknown.

All the poisons we have looked at so far have been obtained from biological sources, many from plants, and most are complex molecules. In the next section we take a look at poisons found in the earth, which are much simpler molecules. In fact, three of them are elements. Despite their simplicity, though, they are no less deadly. Again, as we shall see, such toxins are inherently neither good nor evil; it is the purpose to which they are put that makes them poisons.

PART II

Molecules of Death from the Earth

. { 8 } .

Potassium and the Nightmare Nurse

The potent poison quite o'er-crows my spirit.

—William Shakespeare, *Hamlet*

ESSENTIAL BUT DANGEROUS

How do you commit the perfect murder? First and foremost, you must cleanly dispose of the murder weapon, which can be difficult with a bloody knife or a fingerprint-covered gun. But what if the murder weapon were something simple? What if it simply dissolved into the blood without a trace?

It's hard to imagine that poison would be stocked on the shelves of every grocery store, yet that's exactly where our next poison can be found. Potassium chloride is chemically similar to regular sodium chloride (table salt), and is marketed as a healthier alternative for cooking and seasoning.[1] Potassium is found in all foods with the notable exceptions of vegetable oil and butter; it's vital for the normal function of pretty much every cell in the body. Without it we cannot live—but too much can be lethal.

Interestingly, vegetarians and vegans have higher levels of potassium compared with omnivores, since potassium is particularly abundant in plant foods. Perhaps the best-known dietary source of potassium is the banana. Though bananas are considered to be healthful, the urban legend of death from eating too many bananas and suffering a potassium overdose still lingers. The average banana[2] contains about 450 mg of potassium. Given the recommended daily consumption of 2,500–4,700 mg of potassium, a healthy person can eat at least seven and a half bananas a day before reaching the recommended level. So, can you eat enough bananas to kill yourself? Only if you eat at least four hundred at once.

Unlike other things in the diet, like glucose, fats, or vitamins, which the body can store, there is no storage mechanism for potassium. Thus the body must have a continuous supply to stay healthy.[3] Too little potassium in the body is cause for concern and leads to symptoms of weakness and fatigue, as well as muscle cramps, constipation, and low blood pressure. Low potassium can also depress breathing, reducing the amount of oxygen circulated throughout the body. Very low levels of potassium can affect the normal rhythm of the heart, not only increasing the heart rate but making the beats very erratic and uncoordinated, even leading to heart failure. For this reason hospitals stock concentrated potassium chloride solutions to bring low blood potassium levels back to normal. However, as we will see, too high a blood potassium level is fraught with its own dangers, and sometimes those most trusted to look after us have used their skills not to cure, but to harm.

Kenneth Barlow, the first person convicted of murder using insulin. *Courtesy of True Crime Library*

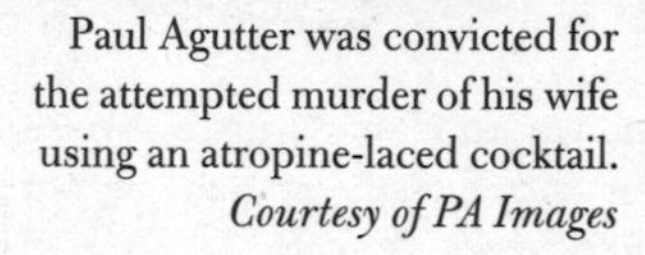

Paul Agutter was convicted for the attempted murder of his wife using an atropine-laced cocktail. *Courtesy of PA Images*

Dr. Cream, a serial killer convicted of murder using strychnine in Illinois and London. *Courtesy of McGill University Archives*

A medical case belonging to Dr. Cream, found with him during his arrest. A bottle of strychnine can be seen seventh from the left on the top row. *Courtesy of the Wellcome Foundation Collection; Science Museum, London*

The gravestone of Daniel Stott bearing the name of his killer, Dr. Cream, in Garden Prairie, Illinois. *Photo by author*

A woodcut of George Lamson's trial for the murder of his brother-in-law using aconite. *Courtesy of the Wellcome Foundation Collection*

Lakhvir Singh, convicted of the murder of Lucky Cheema and the attempted murder of his fiancée by lacing a chicken curry with aconite. *Courtesy of PA Images*

The container of poisoned chicken curry found in Lucky Cheema's fridge. *Courtesy of PA Images*

Georgi Markov, the Bulgarian dissident assassinated via ricin in London. *Courtesy of PA Images*

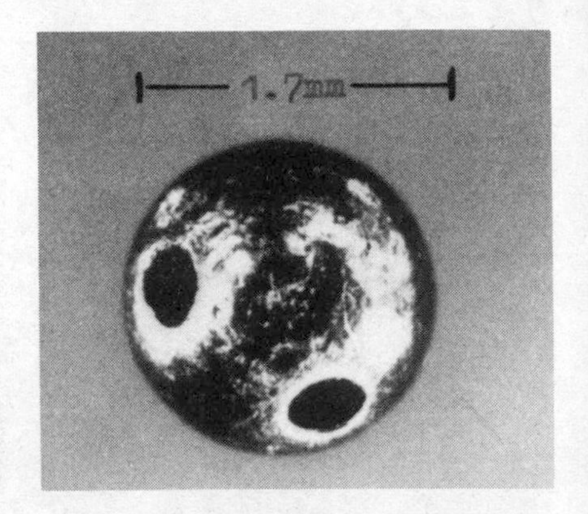

The tiny platinum pellet found in Markov's leg, believed to have contained ricin. *Courtesy of PA Images*

The house where Dr. Robert Ferrante murdered his wife with cyanide taken from his laboratory. *Public domain*

Scientist and killer Dr. Robert Ferrante. *Courtesy of the Allegheny County Prosecutor's Office*

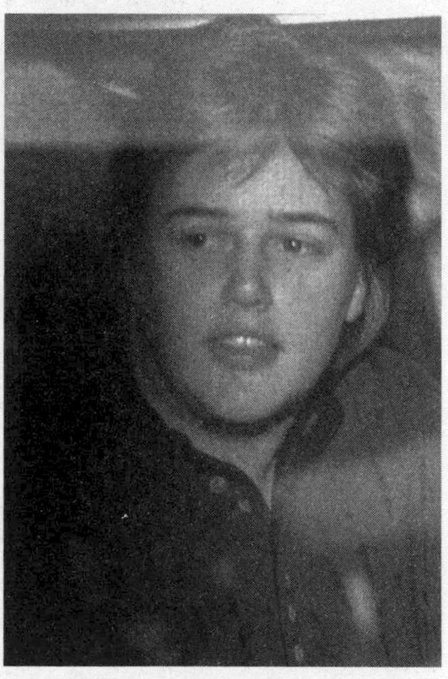

Beverly Allitt, the nurse convicted of murdering young children in her care with an overdose of potassium. *Courtesy of PA Images*

Andrei Lugavoy and Dmitry Kovtun enter the Millennium Hotel in London, preparing to assassinate Alexander Litvinenko. *Courtesy of www.litvinenkoinquiry.org*

The teapot recovered from the Pine Bar in the Millennium Hotel, which contained deadly levels of radioactive polonium. *Courtesy of www.litvinenkoinquiry.org*

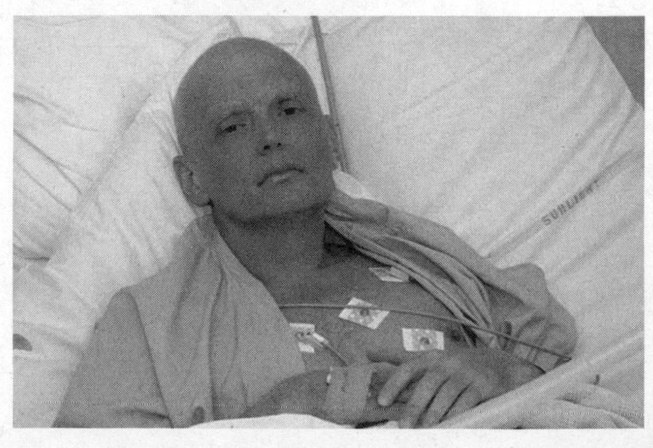

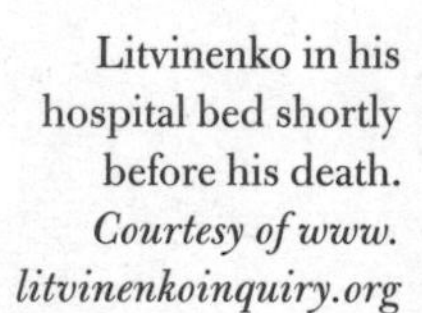

Litvinenko in his hospital bed shortly before his death. *Courtesy of www.litvinenkoinquiry.org*

Madeleine Smith, the Glasgow socialite tried for the murder of her lover with arsenic-laced hot cocoa. *Courtesy of the Wellcome Foundation Collection*

Madeleine Smith's house in Blythswood Square. Miss Smith's bedroom is the second window on the side street. *Courtesy of Glasgow Life*

The bottle of arsenic purchased by Madeleine Smith and presented at trial for the murder of L'Angelier. *Courtesy of the National Records of Scotland*

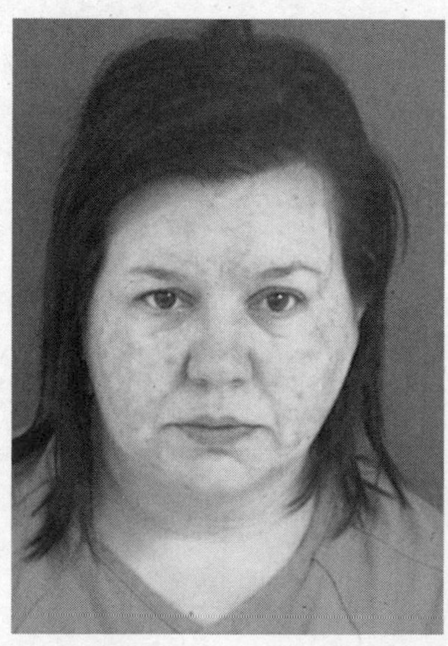

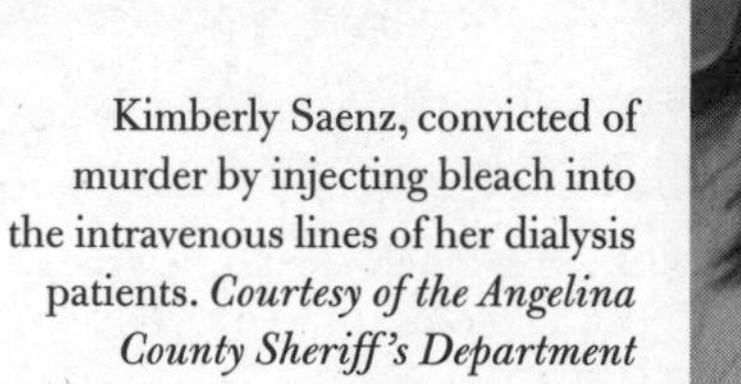

Kimberly Saenz, convicted of murder by injecting bleach into the intravenous lines of her dialysis patients. *Courtesy of the Angelina County Sheriff's Department*

POTASSIUM AND GRANTHAM'S NIGHTMARE NURSE

Low levels of blood potassium can be caused by many things, including excessive alcohol use, uncontrolled diabetes, excessive diarrhea or vomiting, overuse of laxatives, and certain diuretic drugs. Injecting potassium chloride into a patient's blood will quickly raise potassium levels to normal and help stabilize a patient's symptoms.

Potassium injected directly into the blood can be quite irritating, however, and patients with low blood potassium have complained of an incredibly painful burning sensation coursing through their veins as potassium enters the blood. If therapeutic levels of potassium can be painful, one can only imagine the pain of a sudden large influx of potassium into the veins of a young child. One nurse was either unaware or indifferent to the pain she was causing as she injected lethal doses of potassium into the babies and children in her care.

Beverly Allitt was a children's nurse at the Grantham and District General Hospital in the English county of Lincolnshire. Despite needing to take her nursing exams several times before she passed, she was hired into the Ward 4 children's unit due to a staffing shortage. Although her employment lasted only eight and a half weeks, in her brief tenure she managed to poison thirteen children, four of whom died.

Allitt's first victim was seven-month-old Liam Taylor. Liam was admitted to Ward 4 with wheezing and labored breathing from lung congestion. Nurse Allitt comforted Liam's parents, assuring them that their son was in good hands, and that they should leave to get some rest. When Liam's parents returned to the hospital a few hours later, they were told that Liam had taken a sudden turn for the worse and had been rushed into emergency care. Liam's parents asked if they could stay the night at the hospital with their son, and were shown to a special bedroom reserved for parents whose children were seriously ill. Allitt appeared very sympathetic to the family's concerns, even volunteering to work the night shift to look after Liam in case something happened. Around midnight something did happen. Allitt raised the alarm that Liam's heart had suddenly stopped beating. The doctor's strenuous efforts to resuscitate the boy were to no avail. Young Liam had died.

Allitt's next target was eleven-year-old Timothy Hardwick. Timothy, who suffered from cerebral palsy, was admitted to the hospital on March 5 following an epileptic seizure. His parents were impressed with how attentive Allitt appeared toward Timothy, but tragically, when she was left alone with Timothy, his heart stopped beating. Again Allitt called out for help, but it was too late. Nothing could be done to revive him. There was no apparent reason why Timothy's heart would have suddenly stopped, and an autopsy failed to provide any clues. Timothy's death was put down to a complication resulting from his seizure.

Less than a week later one-year-old Kaylee Desmond was admitted to Ward 4 with a bout of lung congestion. Allitt was assigned to the little girl's care, and at first Kaylee seemed to be getting better. Distressingly, however, Kaylee too went into sudden cardiac arrest. Allitt quickly called for the crash team, who managed to revive Kaylee and stabilize her condition to the point where she could be transferred to a larger, better-equipped hospital. Away from Allitt, Kaylee made a full and complete recovery. One thing doctors noticed was a small puncture wound below Kaylee's armpit. Attention was drawn to the area because of a small air bubble just under the skin. Since Kaylee had made a full recovery, this finding was not immediately followed up on, but subsequent police investigation determined that Allitt had injected Kaylee with potassium chloride from a partially filled syringe from which she had failed to expel the air fully. (Little wonder that Allitt had needed multiple retakes of her nursing exams!)

Allitt was apparently frustrated that Kaylee had survived her attack with potassium chloride, and decided her next victim would be injected with insulin. On March 20 five-month-old Paul Crampton came into Ward 4 because of severe bronchitis. Paul seemed to be doing well until suddenly, in the early hours of the morning, he went into a coma. Blood work revealed that Paul's blood sugar level had dropped dangerously low, and a quick injection of glucose seemed to revive him. Paul suffered two further attacks before it was decided that he should be transferred to the larger hospital at Nottingham. Again, away from Allitt's care, her patient made a miraculous recovery.

Allitt returned to potassium chloride for her next victims, five-year-old Bradley Gibson and two-year-old Yik Hung Cha. Both boys suffered cardiac arrests but were stabilized enough to be transferred to Nottingham, where they recovered fully. Unfortunately Allitt's next victims didn't fare as well.

On April 1, 1991, nine-week-old Becky Phillips entered Ward 4 with an upset stomach. Becky had been a premature baby, so her parents were particularly concerned. Becky was promptly examined and diagnosed with a mild bout of gastroenteritis. Treatment was started immediately, and slowly Becky's vomiting and diarrhea subsided to the point where she was ready to return home. During her time at Grantham Hospital, Becky was under the care of Nurse Allitt, who appeared to do all she could to make the little girl's stay as comfortable as possible. She went above and beyond for Becky's family, often staying at the bedside while they went for a break in the cafeteria. After what appeared to be a full recovery, baby Becky was discharged and allowed to go home that afternoon. But not long after arriving home, she started to become restless and show signs of distress, her skin cold and clammy to the touch. Becky's parents rushed her back to the hospital, but it was too late, and she was pronounced dead on arrival.

Worried that Becky's twin sister, Katie, might also have the same illness, her parents brought Katie to the hospital as a precaution. Inexplicably, Katie stopped breathing twice and had to be resuscitated. Fortunately, Nurse Allitt was there each time Katie stopped breathing, and was able to call for help immediately. Although Katie was revived each time she stopped breathing, the lack of oxygen took its toll, and she suffered permanent brain damage. Allitt's quick reactions gave Katie's parents the impression that she was a nurse who was always going the extra mile for her patients. Nurse Allitt had been an angel. The Phillipses thought so highly of her that they even asked Allitt to become Katie's godmother.

On April 22, 1991, fifteen-month-old Claire Peck was admitted to Ward 4 following a severe asthmatic attack. Claire was known to the nursing staff since she had been admitted for several asthmatic attacks during her short life. After each attack the medical staff treated Claire, allowing her to recover sufficiently to be sent back home with her relieved parents. This time was different. Dr. Porter, one of the consultant pediatricians for Ward 4, oversaw Claire's treatment. An asthma attack in someone as young as Claire could seem distressing, but Dr. Porter knew exactly what to do and soon had Claire treated and breathing normally again. He left Claire in the care of Beverly Allitt, one of the pediatric nurses on duty at the time, and went looking for Claire's anxious parents to let them know that all was well.

But all was not well. No sooner had Dr. Porter left the treatment room than the nurse raised the alarm. Little Claire's heart had stopped beating. Dr. Porter was stunned. Claire had appeared to recover fully from her asthma attack; what could possibly have gone wrong in the few minutes since he last saw her?

The resuscitation team responded immediately, and within a few minutes Claire had stabilized. Relieved, Dr. Porter went to tell Claire's parents that the emergency was over, and that their daughter was in no further danger. Once more Allitt was left alone to look after young Claire. Once again, within minutes of his departure, Dr. Porter heard the crash code. Nurse Allitt announced that Claire had once more stopped breathing, and was without a pulse. Dr. Porter ran back to Claire's bed. What had he missed?

Claire lay helpless in the small bed, her lips and cheeks starting to take on a blue hue as her heart and lungs failed to circulate oxygen through her body. The resuscitation team worked hard to try to revive Claire, but this time their efforts were unsuccessful. Claire had gone into cardiac arrest, and her heart could not be restarted. She was pronounced dead within a few hours of being admitted to the hospital. When detectives later questioned Dr. Porter, he recounted that he felt as though something was preventing him from saving the little girl's life.

Claire's parents were devastated by the death of their daughter, and were totally unaware that this was not the first serious incident that had occurred on Ward 4 in recent weeks. In fact, the past few weeks on Ward 4 had seen an unusually high number of its young patients succumbing to tragedy. The number of unexplained deaths among children had now risen to four, and nine other children had collapsed on the same ward.

Following Claire's death, it was finally conceded that a killer was on the loose in Ward 4. But was the killer a staff member, another hospital worker, or someone from outside? To narrow down the suspect pool, a hidden surveillance camera was set up to monitor the entrance to Ward 4. Charting staff schedules and comparing them with adverse medical events on Ward 4, it was found that Beverly Allitt was either on the ward immediately before an episode, or was the person calling for assistance after an episode occurred.

Anyone who has been in the hospital knows that one of the most routine procedures is having blood drawn, and it was no different for the patients on Ward 4. Would the blood samples provide any indication as to what had happened to the Ward 4 victims?

Usually blood samples are supposed to be destroyed within three to six months of their collection. However, an overwhelming amount of paperwork had prevented staff from getting around to discarding many of the collected samples, and there was a chance that the critical evidence of blood samples taken from Allitt's victims was tucked away in the fridge. Remarkably, nine remaining blood samples from thirteen victims were recovered. Among the samples found was blood from the Phillips twins and that of Claire Peck. All had very high levels of potassium, levels consistent with cardiac and respiratory failure.

Allitt was arrested and charged with murder of the children under her care. The trial lasted two months, though Allitt only attended sixteen days due to illness. Despite pleading not guilty to all charges, the jury found Allitt culpable, and she was given thirteen life sentences for murdering four children, attempting to murder another three, and causing grievous bodily harm to a further six. It was the harshest sentence ever given to a female in the UK. In sentencing Allitt to a minimum of thirty years in prison, the presiding judge noted: "I have found that there is an element of sadism in Ms. Allitt's conduct. . . . By her actions, what should have been a place of safety for its patients became not just a place of danger, but if not a killing field, something close to it."

Though her motives are still not fully known, it was argued that Allitt suffered from both Munchausen syndrome and Munchausen syndrome by proxy. Individuals with Munchausen syndrome feign an illness or condition to obtain attention and seek importance. As a child, Allitt often wore Band-Aids on imaginary wounds, careful not to let anyone see the supposed injury. She even went as far as to have a perfectly good appendix removed. Munchausen by proxy, identified by pediatrician Sir Roy Meadow in 1977, occurs when caregivers commit child abuse by deliberately causing, or falsely reporting, illnesses in children to focus attention on themselves. Sufferers from Munchausen by proxy assign another involuntary individual as the patient with the supposed symptoms. In this case Allitt not only caused the patient's distress but was then on hand to "save" the patient through her diligence.

Allitt, the most notorious female serial killer in British history, is serving her sentence at Rampton Hospital, a high-security psychiatric hospital. The impact of Allitt's actions was not only felt by the victims' families, but also by the Grantham Hospital, which shut down its children's unit altogether.

HOW TOO MUCH POTASSIUM KILLS

There are about nine ounces of potassium in our bodies. Almost all of that—more than 90 percent—is tucked away inside our cells, with only a small amount in the blood and fluid bathing the cells. This imbalance between the potassium that is inside and that which is outside is important for every cell in the body, but particularly so for cells that make up nerves and muscles, and especially for heart muscle.

Remove the heart from a person, and that individual will die rather quickly. However, the heart, outside a person, keeps beating quite happily on its own. This is because the heart has its own system for triggering a heartbeat, and doesn't need the rest of the body to do so. On its own outside the body, a human heart will continue to beat at roughly seventy to eighty beats per minute. Although the heart doesn't need the rest of the body to tell it to beat, input from the nervous system can tell the heart to beat slower or to speed up.

Eighty times a minute, special cells at the top of the heart send an electrical signal to the heart muscle cells to contract and squeeze blood to the lungs and the rest of the body. This signal is where potassium enters the story. Muscle cells, including heart muscles, are like tiny batteries: They have a certain voltage and they have positive and negative ends. In the case of heart muscle cells, the voltage is very small, around ninety millivolts. When the heart is resting between beats, the inside of the cell is negative and the outside positive. When heart muscles are stimulated to contract, positively charged sodium ions flood into the cell through their own specific sodium channels. As the amount of sodium inside the cell increases, a small electrical charge is created, causing a brief flipping of the terminals, so that the inside of the cell becomes negative. As the electrical terminals of the heart swap, calcium enters the muscle cell, causing the muscle to contract (the importance of calcium is discussed in detail in chapter 6).

Before another heartbeat can occur, the whole system has to reset back to where it started. Shortly after the sodium channels open, allowing sodium to flood into the cell, potassium channels open to help reverse and reset the polarity. To get the sodium and potassium ions back to their original levels inside the cell, the sodium and potassium channels close, and the sodium pump moves sodium out of the cell, and potassium back in. Although this seems a lengthy process, in reality the whole sequence of events takes less than a fifth of a second. The whole system usually works very well—so well, in fact, that it happens around three billion times in an average life span. But what if something happens to alter this process? What if the amounts of sodium or potassium are suddenly different? What happens if the outside of the heart cell is suddenly bathed in a large amount of potassium, perhaps because someone has just injected a large amount of potassium into the blood?

Think of a person on a train. The train slows as it enters the station, and then stops. The platform is empty, and the passenger has no difficulty exiting the train. Now imagine that same passenger during rush hour. The platform is packed with other commuters, making it extremely difficult to get off the train. Similarly, a lot of potassium already outside the cell makes it difficult for potassium to leave the cell and reset the system. Once the heart muscle has contracted, the failure of potassium to exit the cell means that the heart can't reset and relax. The heart can't beat anymore, and cardiac arrest ensues.

SLAUGHTER IN SHERWOOD

Sherwood is a small town just north of Little Rock, Arkansas. There, on the evening of November 4, 1997, Christina Riggs was tucking her children, two-year-old Shelby and five-year-old

Justin, into bed. But this tender scene was not what it appeared: Christina's actions—far from those of a nurturing mother—were a prelude to cold-blooded murder.

Christina Riggs was born in 1971 in Lawton, Oklahoma, and grew up in a mentally and sexually abusive environment. By the age of the fourteen, she was drinking heavily and smoking tobacco and marijuana. Despite her poor start in life, Christina remarkably not only graduated high school but also attended college to become a licensed practical nurse. Work at the local Veterans Administration hospital and part-time work in a care facility seemed to have settled Christina's life. She even had a steady boyfriend, but when he found out she was pregnant, he wanted nothing to do with the child, and abandoned them both. Her son, Justin, was born in June 1992.

Within a year Christina met a new love, and the couple married. Daughter Shelby was born in December 1994. In 1995 the family moved to Sherwood to be near Christina's mother, so that they would be able to get some help with child care. Christina returned to work as a nurse, now at the Baptist Hospital. However things looked from the outside, the family home was far from idyllic. Christina's husband was unable to cope with Justin's ADHD (attention deficit hyperactivity disorder), once punching the child in the stomach so hard that he required medical attention. The marriage was destined to fail, and the divorce left Christina a single mother with two young children.

Overeating and no exercise caused Christina's weight to balloon to 280 pounds, but she still managed to continue working, and she had enough money to keep her children housed and fed.

But her own depression and a lifetime of bad choices had finally caught up with her, and on November 4, 1997, Christina decided to end it all. She thought she had found the perfect way to end the lives of her children and herself, but the killing did not go as planned. Christina started by giving both her children Elavil, also known as amitriptyline, an antidepressive drug that also has a sedative effect. Once the kids were drowsy and tucked into bed, Christina moved on to the second part of her plan, injecting her children with lethal doses of potassium chloride. Sure that this would be a quick and painless death, Christina did not realize that potassium has to be injected in a particular way to be painless.

Christina injected concentrated potassium chloride straight into the veins of Justin's neck, unaware that potassium had to be diluted in an intravenous drip before entering the blood. Even in a hospital setting, where patients with low blood potassium levels are receiving a slow infusion of fairly dilute potassium by drip, they often report very painful burning sensations as the potassium enters their veins. As the strong potassium solution entered Justin's veins on its way to the heart, the undiluted potassium was destroying Justin's veins. Even though he was sedated, the pain was so severe that Justin was screaming in agony. In a panic Christina reached for another syringe she had stolen from the hospital, this one containing morphine. But again, to be effective, morphine must be injected into a vein. With Justin writhing in pain, finding a vein seemed unlikely and the injection just went under the skin. Desperate, Christine pressed a pillow over his face to muffle his screams, smothering him and cutting off his oxygen supply until he was finally dead. She smothered Shelby without injecting her, then carefully laid her two children side by side on her bed, and tried to take her own life.

To die with her children, Christina had taken twenty-eight of the Elavil pills before injecting herself with the potassium chloride solution. Her first attempt at injecting potassium into her arm failed, as the vein immediately collapsed. Christina was so overweight that she was unable to find another vein, and the potassium chloride she did inject didn't circulate to her heart as she had expected. Nonetheless she collapsed and passed out.

The following day Christina's mother went to the house, but, unable to get in, called the police. When they broke into the house, they found Justin and Shelby lying dead, and Christina unconscious at the foot of the bed. She was rushed to the hospital, where she made a full recovery. Upon discharge she was promptly arrested for the murder of her children. On June 30, 1998, after a mere fifty-five-minute deliberation, Christina Riggs was found guilty of two counts of first-degree murder. The judge sentenced her to death, making her the first woman in 150 years to be executed in Arkansas. It is ironically fitting that executions in Arkansas are carried out by lethal injection—using potassium chloride.

THE RADIOACTIVE BODY

Everyone is radioactive. Every day we eat, drink, and breathe radioactive substances that are naturally found in the environment.

The major source of radiation in the body is a radioactive form of potassium, potassium-40. In a typical adult roughly five thousand atoms of potassium-40 undergo radioactive decay every second. When the potassium-40 decays, it converts either to calcium, a normal constituent of the body, or to the gas argon, which will eventually be exhaled by the lungs.

Although this amount of radioactivity in the body may seem alarming, it is totally normal, and well below what is considered unsafe radiation exposure. Indeed, the deadly effects of potassium on the body are not due to radioactivity, but rather, as we have seen, to the chemical effects of excess potassium on cells. In the next chapter we will come across a poisonous chemical with the opposite properties: one whose chemistry is fairly benign, but whose radioactivity is deadly.

· { 9 } ·

Polonium and Sasha's Indiscriminate Intestine

You may succeed in silencing me, but that
silence comes at a price.
—Alexander Litvinenko, Russian defector, 2006

ARE YOU EATING ENOUGH METAL?

Everyone has heard of the three basic food categories: fats, proteins, and carbohydrates. Depending on the month, or even the week, each of these in turn is demonized or praised in the media. However, all three components are required for a healthy body. What is less well appreciated is the need for metal in our diet. Although chemically things like sodium, potassium, and calcium are classified as metals, we tend to think of metals as—well—metallic, like iron or copper or aluminum. Nonetheless, metals play key roles in the human body, important in breathing, fighting off infections, and making strong bones. Iron is a vital component of every human body, central to the ability of blood to carry oxygen around the body. Copper is found in a healthy immune system, as is zinc. Manganese, found in all cell phones, also plays a pivotal role in brain function. Given the importance of metals in the body, it is not surprising that humans have specialized mechanisms for absorbing metals from our diet.

Although some metals are necessary for normal function, others, like lead, cadmium, and polonium, are lethal. Fortunately humans rarely encounter these metals, as they are usually found as compounds or minerals located deep underground. However, the advent of mining and smelting has brought these metals into the environment, where they now have more potential to enter the body.

A SHORT HISTORY OF POLONIUM

In 1903 Pierre and Marie Curie won a Nobel Prize in Physics for their work on radioactivity and the discovery of a new radioactive element, which they named as a tribute to Poland, the country of Marie's birth: polonium. Sadly, the first casualty of polonium's radioactivity was Irène Joliot-Curie, daughter of Marie and Pierre, who died of leukemia in 1956 possibly due to an accidental exposure to the volatile metal.

Polonium is actually very rare, only present at 100 micrograms (100 millionths of a gram) per ton of ore. In the 1920s physicists discovered that they could make new elements by bombarding known elements with radiation. This set off a frenzy of physicists giddily irradiating anything they could get their hands on to make new, previously undiscovered elements. Eventually scientists would achieve the alchemist's dream by bombarding lead with radiation and converting it into gold, but at a cost that far exceeds the value of the gold generated. It was also found that irradiating the element bismuth resulted in the generation of polonium-210. In the 1950s and 1960s tests with animals showed just how dangerous polonium-210 was, revealing that only a microgram the size of a particle of dust, was lethal.

Polonium-210 found use as a trigger for nuclear weapons, and at one point the United States, the USSR, Britain, and France all had nuclear reactors making polonium for their bombs.[1] When scientists discovered that nuclear weapons could be set off much more efficiently with tritium (a radioactive isotope of hydrogen) the NATO nuclear countries shut down their polonium production, leaving the new Russian Federation as the only country producing polonium-210. The Mayak nuclear reactor, located east of the Ural Mountains near the city of Chelyabinsk, is now responsible for the entire worldwide supply of polonium.[2]

Polonium-210 appears to be the ideal assassin's tool. It is lethal in small amounts, being 250,000 times more deadly than the same weight of cyanide. It also doesn't generate the hard gamma radiation that is easily detected by monitors at airports and shipping ports. Although death from polonium radiation is quick, it isn't instantaneous, giving time for the assassin to escape before death strikes.

Is polonium-210 the perfect murder weapon? Well, the reader can decide, as a case in London at the end of 2006 unfolded in the manner of a bestselling Cold War crime thriller.

THE CASE OF EDWIN CARTER

Edwin Carter came home feeling unwell. Maybe it was a cold coming on, or something he had eaten. At 11:00 p.m., Edwin and his wife went to bed, but barely ten minutes passed before he was retching and vomiting. After an hour Edwin felt a little better but decided to spend the rest of the night in the study so as not to disturb his wife and son. After a night of vomiting, he was exhausted. He had stomach cramps and difficulty breathing. He spent the next day at home resting, while his wife begged him to let her call for an ambulance. At first Edwin was reluctant to call the paramedics, but at 2:00 a.m. the next morning he relented.

The ambulance took Edwin to Barnet and Chase Farm Hospital in North London. The diagnosis at the hospital was gastroenteritis with dehydration. While this seemed an obvious answer when presented with vomiting and diarrhea, Carter's white cell count argued against it. Typically, when a patient has an infection, the white blood cell count goes up, since white cells are part of the body's immune system to help fight off infection. However, anticipating finding a high white cell count in Edwin's blood, the doctors were surprised to find his white cell count to be extremely low.

Despite numerous tests, doctors still had no idea what was wrong with Edwin Carter. He was clearly in pain, his body undergoing recurrent episodes of diarrhea and vomiting. Raw ulcers peppered the patient's throat, making it painful for him to eat or drink. Initially the medical staff put him on the broad-spectrum antibiotic ciprofloxacin. Although the doctors were stumped, Carter claimed he knew what was wrong with him. He also announced that he was an ex-KGB agent and had been poisoned by the heavy metal thallium.

Hospital staff were unsure if the patient was delusional or if his infection was affecting his brain. A week after being admitted to the hospital, something strange started to happen to Edwin Carter—his hair was falling out. Despite the medical staff's conviction that he was suffering from gastroenteritis, certain symptoms didn't fit. The sudden loss of his hair (a process called alopecia), along with an almost nonexistent platelet count, was certainly not consistent with gastroenteritis, but neither was any other known disease. Carter continued to maintain that he had been poisoned. While this seemed unlikely, toxicologists at Barnet Hospital agreed to test for heavy metals. Samples were sent off for analysis, and remarkably came back positive for thallium.

A provisional diagnosis of heavy metal poisoning with thallium was made, but although detectable, the levels of thallium were barely above those found in the environment. Nonetheless this new diagnosis caused two outcomes. First Scotland Yard was notified, and second, Carter was put on the only known treatment for thallium poisoning, Prussian blue (see chapter 7 on cyanide). The police arrived just after midnight and started interviewing Carter.

He began with a shocking claim: He told the police that his real name was not Edwin Carter but Alexander Litvinenko, and that he was a former lieutenant colonel in a top-secret department of the KGB. Carter offered one piece of evidence in support of his strange tale: a telephone number. When the number was called, a man, identified only as Martin, answered and agreed to come to the hospital. Martin, an MI6 officer, confirmed that Carter was Alexander Litvinenko, a former KGB agent who had defected and was now advising MI6 on Russian organized crime.

Alexander Valterovich Litvinenko, known as Sasha, was born on December 12, 1962, in the Russian city of Voronezh, some three hundred miles south of Moscow. Following in his grandfather's footsteps, Sasha joined the army, rising to the rank of platoon commander. In 1988 Litvinenko was transferred to Moscow to a special division of the Ministry of Internal Affairs, where he was recruited into the KGB. Litvinenko began his "spy career" in military counterintelligence before joining a department focused on combating organized crime, corruption, and terrorism. The day after Christmas in 1991, the Soviet Union ceased to exist, and with it the KGB. Litvinenko's old KGB unit was rolled into a new service, the Federal Security Service of the Russian Federation (FSB), where he continued his work against organized crime. Following the collapse of the USSR, the Russian economy went from a communist command economy to a capitalist free-for-all almost overnight. These conditions were perfect for the establishment of "crime-bosses" who turned Russia into a version of 1920s Chicago.

Litvinenko became disillusioned with the whole system when he found that his own superiors were in league with organized crime, and that corruption was rampant. He had found evidence of an organized crime syndicate trafficking heroin from Afghanistan to Western Europe, a syndicate he was convinced was colluding with FSB officials, including Vladimir Putin. In the eyes of colleagues in the intelligence services, Litvinenko then committed a cardinal sin by holding a press conference and publicly airing the FSB's dirty laundry. "The FSB infrastructure has been used by certain officials not for the constitutional aims of security of the state and the individual, but for their own political and financial gain," he told reporters.[3] Clearly his FSB bosses were not pleased with his exposé, arresting him on trumped-up charges and jailing him for several months.

It didn't help matters that Litvinenko had alleged that Putin was personally involved in the cover-up of organized crime activities connected with drug trafficking in Russia and Europe. In response Putin held a television interview in which criticism of the messenger was meant to impugn the message, stating: "FSB officers should not stage press conferences and should not expose internal scandals to the public." In January 1999 Putin fired Litvinenko from his job at the FSB. Without a job, and with serious concerns for his family's safety, Litvinenko took the bold step of defecting to the West. The British government gave Litvinenko a British passport, an encrypted phone, and a salary of £2,000 per month. Litvinenko was now an informant for MI6.

THE INDISCRIMINATE INTESTINE

The average adult has around twenty-eight feet of intestines carefully folded into their abdomen. The part responsible for digestion and absorption of food is the small intestine, twenty-three feet of coiled tissue that sits between the stomach and large bowel. It is here that digestive enzymes go to work on food, breaking it down so that it can be readily absorbed. The intestine is lined with a single layer of cells called the intestinal epithelium, which transports nutrients from inside the intestine to the blood. As with the transport of any goods, specialized carriers are needed to move things along efficiently. It is also important that different kinds of transport are required for different cargo, each specialized for the cargo it carries. Sugars, amino acids, and fats all use different transport proteins in the epithelium to get into the body.

Metals also use their own transporters, and substances like iron and zinc are moved into the cells of the intestine using a special transport protein called DMT1. DMT1 doesn't discriminate between iron, copper, and zinc, happily bringing these metals into the body. However, DMT1 cannot tell the difference between metals the body needs and dangerous metals like lead, cadmium, and polonium. When confronted with metals that are not good for the body, the DMT1 transport system imports the deadly metals into the body anyway.

MURDER IN MAYFAIR

Because of his connections in Moscow and his understanding of Russian business practices, MI6 introduced Litvinenko to Titon International, a commercial intelligence company that helped businesses interested in expanding into markets like the former

Soviet Union. In 2005 Litvinenko received a phone call, later followed up by a dinner meeting, from Andrei Lugavoy, a successful Moscow businessman. Lugavoy offered a partnership: Litvinenko would find London-based companies seeking to do business in Russia, and Lugavoy would perform due diligence and collect commercial information on the Russian companies involved. It was not surprising then, that when Lugavoy called to say he would be in London in November 2006, Litvinenko readily agreed to meet him.

The Millennium Hotel sits on the southern side of Grosvenor Square, in London's upscale Mayfair district. The U.S. Embassy used to be on the western side of the square,[4] flanked by statues of Presidents Dwight D. Eisenhower and Ronald Reagan. An inscription praises Reagan's intervention to bring about the end of the Cold War and the dismantling of the Soviet empire. A tribute from Mikhail Gorbachev reads: "With President Reagan, we traveled the world from confrontation to cooperation." In an ironic twist, the Millennium Hotel, a stone's throw from where Reagan's statue stood, was the site of the assassination of former KGB agent Alexander Litvinenko.

Shortly after four o'clock on the afternoon of Wednesday, November 1, 2006, two Russians, Andrei Lugavoy and his business partner, Dmitry Kovtun, entered the Pine Bar in the Millennium Hotel. Ostensibly the Russians were there with their families to watch a critical international soccer match between London's Arsenal team and CSKA Moscow. As the pair sat down, a waiter came to ask what the men wanted to drink. By a striking coincidence, the waiter, who had worked at the Pine Bar for more than twenty-five years, had served many celebrities. Among these was the late Sean Connery, the first and best-known cinematic portrayer of British spy James Bond. As Bond, Connery had been responsible for thwarting many Russian plots, but today the Russians would succeed.

A pot of tea was ordered, and at 4:30 p.m. Litvinenko entered the Pine Bar and joined Lugavoy and Kovtun. Cups of tea had already been poured. Lugavoy asked the waiter to bring a new cup for Litvinenko. There was barely any tea left in the pot, and even that had started to get cold, but Litvinenko still took a few sips. Those sips were to seal his fate. Though he didn't yet know it, Litvinenko's body was already starting to fall apart.

AN ALMOST UNDETECTABLE POISON

Although everyone at the hospital was now convinced that Litvinenko had been poisoned, no one was sure with what. The assumption had been thallium, but the Prussian blue treatment seemed to be having little effect, indicating that thallium was not involved. Could it be another heavy metal? Tox screens came back negative for other common heavy metal poisons.

Then one of Litvinenko's physicians noticed the similarity between him and leukemia patients undergoing chemotherapy. Could he have been overdosed with chemotherapy drugs? The idea of radiation was also suggested, and Litvinenko's body was scanned with a Geiger counter. The results—nothing detected. However, Geiger counters detect only harsh gamma radiation.

The hospital didn't have the equipment to detect the much rarer form of radiation, alpha radiation. Only the British nuclear weapons center at Aldermaston had that capability.

A liter of Litvinenko's urine was sent to Aldermaston, but the test would take more than twenty-four hours. Meanwhile, Litvinenko's life was slowly fading away as he slid in and out of consciousness. Litvinenko's heart was becoming weaker, and on the night of November 22 he went into cardiac arrest. The crash team rushed into action, but it took thirty minutes to resuscitate Litvinenko. The next afternoon Aldermaston called with the results from Litvinenko's urine: The poison was finally identified as polonium-210. The amount was a million times the lethal dose. Litvinenko was living on borrowed time.

It would later be found that Litvinenko had an incredible 26.5 micrograms (26 millionths of a gram) of polonium in his blood. Despite this minute amount of poison, the radiation attacking his body was equivalent to having 175,000 X-ray images taken. Less than one microgram would have been more than enough to cause death. Part of the reason it had taken so long to figure out that Litvinenko had been poisoned with polonium was the fact that it had never been used before as a murder weapon.

Three weeks after visiting the Millennium Pine Bar, Litvinenko suffered another cardiac arrest. Twenty-one minutes later he was pronounced dead, and his hospital room was sealed.

Eight days after Litvinenko's death, pathologists examined his body. The postmortem was one of the most dangerous ever undertaken in the West. The people involved in the autopsy would not have looked much different had they been dissecting an alien in Area 51, a top secret USAF facility in Nevada much favored by conspiracists. The forensic pathologist, Nathaniel Cary, wore two protective suits, his gloves taped at the wrists, and filtered air pumped into his plastic hood. A second pathologist, a police detective, and a photographer were all similarly dressed. A radiation specialist stood by to wipe contaminated blood off any of the people at the autopsy. Paramedics were waiting to instantly evacuate anyone who showed the slightest sign of falling ill.

When the corpse was opened, all that could be seen was atrophied and decayed tissue, shredded and dissolved from the inside out. Polonium's radiation had relentlessly destroyed Litvinenko's body.

RADIATION AND THE INTESTINE

The daily strain of digestion and absorption takes a toll on the cells lining the intestine, and as they die they are sloughed off from the surface, like skin cells after a sunburn. These dead cells are digested and recycled back into the body to make new cells. The whole process goes on continuously and automatically. In fact, all the cells lining the small intestine are replaced roughly once every three to seven days, making the intestine the fastest-replicating tissue in the body. This rapid growth and replication of cells in the gut requires lots of DNA synthesis, and although this process is incredibly efficient, its speed also makes the intestine very sensitive to substances that interfere with DNA.

Inside each cell is a nucleus that houses the genome, which

contains all the instructions needed to make new cells. As Matt Ridley puts it in his book *Genome: The Autobiography of a Species*: "Imagine the genome is a book, there are 23 chapters called CHROMOSOMES. Each chapter contains several thousand stories called GENES. Each chapter is made up of paragraphs called EXONS, which are interrupted by advertisements called INTRONS. Each paragraph is made up of words called CODONS, and each word is written in letters called BASES." There are roughly three billion letters in the book—equivalent to 250,000 Bibles—yet it all fits into a structure smaller than a pinhead. Each cell in the body contains 6.5 feet of DNA, all tightly coiled and packaged inside a nucleus that is only 6 microns (6 millionths of a meter) wide.

The DNA itself is made up of a four-letter code, A, T, G, C, arranged in various triplet combinations that act as the blueprint for all the proteins a cell will need, whether it be a muscle, heart, brain, or intestinal cell. Each time a cell divides, these three billion letters must be copied exactly, with no errors. Despite the enormity of this task, it takes a cell roughly an hour to replicate its entire three-billion-letter DNA sequence. In contrast, a medieval monk, working fourteen hours a day, usually took four years to copy the roughly three million letters in the Bible.

Sometimes minor errors do occur, and the cell has repair mechanisms to compensate; large-scale damage to DNA, however, cannot be repaired. Cells that divide only occasionally are less prone to the consequences of damage to their DNA. However, cells that divide very rapidly, like intestinal cells, or cells of the immune system, are very sensitive to anything that would break the strands of DNA in the nucleus. One of the things that can destroy DNA strands is radiation, and it can make the DNA chains impossible to repair.

Polonium-210 emits a kind of radiation called alpha-particle radiation. For the most part alpha particles are innocuous and can be easily blocked by a sheet of paper or even the skin coating our bodies, and rarely present any danger if we are exposed. If the radiation is ingested, however, it's a different story altogether. Although the exact form of polonium that killed Litvinenko was never revealed, it was likely ingested as polonium chloride. Although polonium is a solid metal at room temperature, it can be converted into polonium chloride making the polonium soluble in water and much more easily absorbed. Not recognizing polonium as a lethal metal, the intestine's DMT1 is unable to discriminate, and rather than bar polonium from entering the cell, it eagerly brings polonium in, oblivious to the danger it presents.

Inside the cell the alpha particles, emitted by the disintegrating polonium, act in the same way as a wrecking ball demolishing a house. DNA strands are irreparably smashed into tiny fragments; cargo trafficking proteins are blasted apart; and cells have no defense mechanism against this radioactive onslaught. Alpha particles also impact another component of all cells: water. Like a right hook from a bare-knuckle prizefighter hitting the jaw and sending a tooth flying, alpha particles smash into water molecules, knocking off an electron. The sudden loss of an electron causes free-radical oxidants to form. These highly reactive molecules careen through cells smashing important chemical bonds in proteins, cell membranes, and DNA. As the polonium-210 breaks down the gut wall, bacterial infections erupt, leading to peritonitis and causing toxic shock syndrome, a major medical problem on its own.

From the gut, polonium-210 enters the bloodstream, where its first stop is the liver. Here the alpha particles indiscriminately destroy the liver cells, like Vandals sacking Rome. One of the functions of the liver is to help the body clear waste products that arise from the breakdown of old red blood cells. As these discarded red cells break down, hemoglobin is released and further broken down into a compound called biliverdin. In healthy individuals the liver rapidly recycles the components of biliverdin, but in a damaged liver biliverdin accumulates, leading to the characteristic yellow-green skin pallor of jaundice. Moving out of the liver, polonium-210 enters the heart, where the muscle is shredded by alpha particles, eventually causing heart failure. Other cells in the body that have a high rate of division, such as hair follicles, are also torn apart, leading to rapid hair loss.

Finally the deadly radiation attacks the cells of the immune system, killing off the white blood cells that would otherwise protect the body from infections. The source of white blood cells in the body is the bone marrow. Here cells called stem cells rapidly divide and proliferate, maturing into all the various cells found in the bloodstream, such as white and red cells. As we have seen before, rapidly dividing cells are a vulnerable target for radiation damage, and as the chaos in the immune system continues, the number of white cells in the blood drops drastically. The bone marrow is also the source of platelets, the tiny cells responsible for clotting the blood. With the bone marrow damaged, platelet counts drop, clotting stops, and blood loss from internal bleeding ensues. All these symptoms were experienced by Litvinenko as his body was literally being shredded apart, ultimately leading to his death.

WHO KILLED LITVINENKO?

The cost of the polonium used on Litvinenko, had it been bought on the open market, would have been tens of millions of dollars. Such an expensive method of assassination would not be available to a disgruntled individual, or even the criminals of the Russian mob. However, it could be easily available to a state-sponsored organization. The only source of polonium is a nuclear reactor. Every batch of polonium-210 that is made has its own chemical signature, equivalent to a chemical fingerprint, which reveals where it was made. The polonium used to kill Litvinenko had been generated at the Mayak nuclear facility in Russia and brought to London in October, aboard a flight from Moscow.

There is ample evidence that the people responsible for putting polonium-210 in Litvinenko's tea were Andrei Lugavoy and Dmitry Kovtun. It is not clear if either had a personal animosity toward Litvinenko, and they were likely just following orders from higher up. Their cavalier attitude in handling the polonium-210 suggests that they were completely unaware of how dangerous it was. In fact, Lugavoy even encouraged his eight-year-old son, who was traveling with him, to shake Litvinenko's hand after he had drunk the polonium. Wherever Lugavoy and Kovtun went, whatever they touched or sat on, all gave a telltale signal of alpha radiation, allowing detectives to map the pair's movements precisely.

When scientists entered Lugavoy's hotel room, it was like entering a nuclear reactor. The living room of the suite had radiation readings of more than thirty thousand counts per second. The bathroom was even worse, with radiation levels so high that the equipment could not count it fast enough.

While it is generally acknowledged that Lugavoy and Kovtun were the foot soldiers, the identity of the person or persons giving the order for Litvinenko's assassination is still mired in conspiracy. Litvinenko himself was convinced that the order came directly from Vladimir Putin. Whether this belief stems from knowledge or hubris is not known. Was Litvinenko really that important for Putin to be involved? Certainly, there was a personal dimension to the antagonism between Litvinenko and Putin, which may have played a part in the alleged assassination. Much of the evidence in the British government's report on the killing is still highly classified, but the report does state "that in general terms, members of the Putin administration, including the president himself and the FSB, had motives for taking action against Mr. Litvinenko, including killing him."[5] Litvinenko's wife and brother are more inclined to believe it was senior members of the FSB, who believed that Litvinenko was a traitor for exposing any actions of the FSB, and had to be dealt with as an example for others who might be considering similar revelations. Despite assertions of the Russian government's involvement in Litvinenko's death, officials in Moscow have always strenuously denied any involvement in the killing, or in any black market trading of polonium. In May 2007 the UK's Crown Prosecution Service formally charged Andrei Lugavoy with Litvinenko's murder; Putin refused to extradite Lugavoy to the UK. Lugavoy protested his innocence, and held a press conference to criticize what he called the unwarranted and fabricated evidence against him. The press conference was held in the exact same room that Litvinenko had denounced government corruption when he was still with Russian intelligence.

As far as is known, Alexander Litvinenko is the only person to be murdered with polonium-210. The fact that the poison simply did not exist before the nuclear age and its high price of production both probably play into this statistic. While polonium-210 may have had a short-lived existence as a poison, the next chapter highlights a poison that has been known and used since the days of ancient Rome.

.{ 10 }.

Arsenic and Monsieur L'Angelier's Cocoa

They put arsenic in his meat / And stared aghast to watch him eat.

—A. E. Housman, *A Shropshire Lad,* 1896

A SHORT HISTORY OF ARSENIC

Arsenic likely holds the record of having the longest and most disreputable pedigree as a poison. Suggested as the reason for Alexander the Great's demise, considered by Cleopatra for her suicide, and responsible for the rise of Nero to the throne of Rome, arsenic has been a slayer and maker of rulers since ancient times. Indeed, the word *arsenic* comes from the Greek *arsenikos*, meaning "virile" or "masculine."

In Renaissance Europe, the most notorious practitioners of arsenic poisonings were the Borgias, headed by the Spanish-born cardinal Rodrigo Borgia, who poisoned his way through the hierarchy of the Roman Catholic Church to become its head as Pope Alexander VI. Along with his son Cesare and daughter,

Lucrezia, he experimented with various arsenic poisons, trying to find the most efficacious. One recipe involved smearing arsenic on the entrails of a dead pig, which was then left to rot. The whole mess was dried and made into a powder, and along with other secret components made cantarella, a poison that, according to legend, was so deadly that the formula was destroyed after the Borgias died.

As pope, Rodrigo could appoint cardinals within the church. Becoming a cardinal was extremely lucrative, since an appointee could increase his personal wealth through the sale of indulgences, a payment to the cardinal to forgive in advance sins that the purchaser was often planning on committing as soon as he or she left the church building. When each cardinal had acquired a substantial fortune, he was invited to a sumptuous banquet hosted by the Borgias. During the evening the unsuspecting cardinal would be plied with wine heavily tainted with cantarella. Naturally everyone would be shocked and saddened by the cardinal's untimely death, and according to Church rules, all the deceased cardinal's wealth and property reverted to the church—that is, it all went to the Borgias.

So diligent and skilled was the Borgia crime family that it became one of the richest families in all of Italy. Their status was further enhanced by Lucrezia's three marriages into moneyed families, and by Cesare's position as captain-general within the papal army. However, the Borgia dynasty was soon to collapse. On an occasion when several cardinals were due to dine with the pope and his family, Rodrigo and Cesare arrived home early and called for a bottle of wine. Whether by accident or design, a servant poured from a bottle laced with arsenic. The aging pope died, but the younger Cesare, sensing that they had been poisoned, called for a mule to be slaughtered and dressed—a well-known therapy for poisoning at the time was wrapping oneself in an animal carcass. Cesare's recovery is perhaps the only documented proof that the remedy worked, but with his father gone, Cesare never again had the wealth and power he desired. He would die in a minor skirmish in 1507 at the age of thirty-one. Lucrezia fared somewhat better, apparently repenting of her murderous lifestyle and giving her life to religious devotion. The popularity of arsenic poisoning, however, continued for centuries after the demise of the Borgias.

In France in the late 1600s, the effectiveness of arsenic in disposing of wealthy relatives who had the temerity to remain alive was so widespread that it gained the name *poudre de succession*, or "inheritance powder."[1] Extracting arsenic from its mineral deposits was a difficult, time-consuming process, which made arsenic expensive—and murders with arsenic the province of the wealthy. But that changed with the Industrial Revolution, and its huge demand for iron and lead. These metals were extracted from mineral ores that were often contaminated with arsenic. To obtain the pure metal, the ores were heated to high temperatures in large kilns until the molten metals ran free. The arsenic reacted with oxygen to form arsenic trioxide, which condensed as a white powder and was periodically scraped from the kiln's chimneys so that they didn't become clogged.

Rather than throw the "white arsenic" away, it was realized that money could be made by selling the substance as a poison to get rid of all kinds of vermin, including cockroaches, rats, stray animals—as well as relatives and secret lovers. Now that arsenic was being produced on an industrial scale, its cost plummeted, and even the poorest person could obtain it to deal with an unwanted problem. In 1851, with increasing public concern over accidental and deliberate arsenic poisonings, the British Parliament passed the Arsenic Act, which sought to regulate who could buy arsenic.[2]

In the same way that cyanide formed the basis for Prussian blue, Carl Wilhelm Scheele discovered he could use arsenic to make a brilliant green color, later named Scheele's green in his honor, that became all the rage in things as diverse as clothing, wallpaper, candy decorations, children's toys, and soap. Even the German chemist Robert Bunsen (later of "Bunsen burner" fame) wanted in on the arsenic craze. As he was sloshing around arsenic compounds one day, the explosion of the glass beaker nearly took out his right eye, leaving him half blind for the rest of his life.

One of the great appeals of arsenic to would-be poisoners was that physicians often mistook symptoms of arsenic poisoning for natural illnesses. This was especially true if the victim died from the accumulation of many small doses over time. Often poisoning was mistaken for cholera, influenza, or even simple food poisoning, conditions that were common parts of life before the twentieth century. There is no way of knowing how many murders were simply ascribed to disease.

The initial symptoms of acute arsenic poisoning are gastrointestinal upset, severe vomiting, and diarrhea. With so much loss of fluids, dehydration and intense thirst set in, along with extreme stomach pain. The corpses of acute arsenic poisoning victims can appear slightly shrunken or emaciated due to rapid and severe dehydration. The vomiting and diarrhea are caused by irritation of the stomach lining, which can be seen on autopsy as bloodied lesions in the stomach lining. Arsenic can also attack the intestines, where similar damage can be found.

But arsenic doesn't kill only by acute poisoning in one large dose. It is also deadly when a victim is fed small amounts of arsenic over a long period, with the arsenic slowly accumulating in the body—this is called chronic poisoning. Slow chronic arsenic poisoning is the preferred method of killing by those who wish to mimic natural disease. Many arsenic poisoners were very attentive nurses or spouses, who were always on hand to administer repetitive doses until the desired result was achieved. At these lower doses, symptoms of vomiting and diarrhea do occur, as well as headaches, nausea, and dizziness. Muscle cramps and paralysis are also common due to accumulated nerve damage, along with an irregular heart rhythm that can send the pulse racing. Victims can live with these symptoms for many weeks before finally dying of multiple organ failure. A common feature of chronic arsenic poisoning is the appearance of dark spots on the skin (hyperpigmentation) and even the formation of hard scaly patches called arsenical keratoses. An examination of the fingernails would reveal "Mees lines," longitudinal white lines on fingernails that run parallel to the nail bed.

Arsenic has two things going for it to make it a popular poison. First, it is very soluble, and second, in contrast to many of the plant alkaloid poisons, it has barely any taste. This makes it easy to sprinkle on food or stir into soups and stews. However, the most common approach is to dissolve the arsenic in something the victim would normally drink, such as wine, coffee, or cocoa.

In this form a mere mouthful of a poisoned drink might be enough to kill. However, some people—including the population of one alpine village in Eastern Europe—have survived the ingestion of levels of arsenic that would be lethal to others.

THE ARSENIC EATERS

Located near Austria's regional capital of Graz, on the border with Hungary, lies the region of Styria. One of the more notable sons of Styria is Arnold Schwarzenegger, champion bodybuilder, film actor, and former governor of California. Roughly one hundred years before Schwarzenegger was born, a surprising report by the Swiss naturalist Dr. Johann Jakob von Tschudi appeared in an 1851 Viennese medical journal, describing peasants in the alpine region of Styria who regularly ate arsenic as a tonic.

The peasants would crunch lumps of white arsenic between their teeth or grate it onto bread, two or three times a week. The men claimed that the arsenic helped them breathe better at the high altitude of the Styrian Alps, gave them more physical bulk, aided digestion, prevented disease, and increased their sexual potency. For their part the women argued that eating arsenic greatly improved their skin tone, giving them a "peaches-and-cream" complexion as well as a more curvaceous figure. Arsenic does stimulate the production of hemoglobin—and red blood cells—which can increase oxygen transport through the blood and may provide a clue as to the Styrians' claim that arsenic helped them breathe better at higher altitudes.

The Styrians first gained a taste for arsenic in the 1600s, when mining started in the region. When minerals containing arsenic were smelted, arsenic trioxide deposited as a white powder in the chimneys above the smelting fires. The arsenic powder was collected, and like salt, sprinkled on bread or dissolved in a warm liquid such as coffee. Why the miners started doing this, and where the idea came from, is still shrouded in mystery. The young would start with a small amount of arsenic, the size of a grain of rice, and gradually increase the dose until they could eat quantities normally considered lethal with no apparent ill effects. Indeed, as a rule, arsenic eaters generally lived long lives free from many infectious diseases, having eaten arsenic regularly for more than forty years. Many of the men routinely consumed 300 mg of arsenic, well above the lethal dose for an adult; one individual was reputed to eat almost a gram of white arsenic on a regular basis. Not only did arsenic apparently improve the lives of the men and women of the region but the powder was also fed to their horses. Incredibly, they claimed it improved their animals' health and appearance, and increased their stamina, too.

In fact, arsenic is an essential trace nutrient for many animals. Studies have shown that tiny amounts of arsenic fed to chickens stimulates the formation of blood vessels, making the chickens appear plump and nicely pink. As late as 2013, all U.S.

chickens were fed arsenic in their diets. As yet it is not known if arsenic is essential for humans, but it could quite possibly help increase the body's blood supply and improve stamina at high altitudes.

Since the scientific and medical community of the mid 1800s was well aware of the deadliness of arsenic, the idea that people could eat arsenic with impunity seemed to be a legend on a par with Bigfoot or the Loch Ness monster. A public scientific demonstration seemed appropriate to dispel disbelief in this new discovery. At the forty-eighth meeting of the German Association of Arts and Sciences at Graz in 1875, two arsenic eaters were introduced to the audience, one of whom ate 400 mg of white arsenic, and the other 300 mg. The next day the two men were again introduced to the audience, both in perfect health. Moreover, samples of urine taken from each man demonstrated an abundant presence of arsenic. There could be no more doubt that it was possible to eat arsenic and develop an apparent immunity by gradually increasing the dose.

After death, one of the more bizarre results of eating arsenic is that the poison kills off the bacteria that would normally lead to putrefaction of the corpse. The burial traditions of the Styrians involved taking the corpse from the grave after twelve years so that the bones could be removed and placed in a crypt, leaving the grave free for a new occupant. The bodies of the arsenic eaters were often so well preserved that, even after twelve years the disinterred corpses were easily recognizable to friends and family. The ability of arsenic to dramatically slow down decomposition after death is referred to by toxicologists as "arsenic mummification." It has been proposed that the origins of the vampire legends of the undead, which arose in Central and Eastern Europe, may in part lie with the well-preserved bodies of deceased arsenic eaters.

The publicity given to the supposed health benefits of arsenic eating helped popularize the use of arsenic in medicines and cosmetics. Not only was the general public enthralled by taking arsenic but the legal community also seized upon its usefulness.

Lawyers who were defending clients accused of committing murder by arsenic poisoning came up with what was to be called the Styrian defense.[3] The theory contended that arsenic found in a dead body was not evidence of foul play but a sign that the victim was an arsenic eater who took the powder as a tonic. If the poisoning was self-inflicted, no crime had been committed, and therefore the accused had to be let go. Similarly, finding arsenic in the possession of an accused was not de facto evidence of malice, because she—and often the accused poisoner *was* a she—could have been applying the arsenic to her skin to improve her complexion.

The Styrian defense was a boon to defense lawyers and was used in many trials, including, as we shall see, the trial of the Glasgow socialite Madeleine Smith. One skeptical correspondent in *Chambers's Edinburgh Journal* wrote: "Let me urge upon all who adopt the Styrian system [eating arsenic], to make some written memorandum that they have done so, lest in case of accident, some of their friends may be hanged in mistake." A cautionary word indeed!

ARSENIC AND MONSIEUR L'ANGELIER'S COCOA

High society, scandal, Victorian sensibilities, blackmail, and murder: What more could any journalist want? Newspapers called it the "Trial of the Century" and a "thrilling narrative of crime, passion and judicial inquiry." On Thursday, July 9, 1857, the atmosphere outside the High Court in Edinburgh was electric as crowds awaited the jury's verdict. At stake was the life of accused murderess Madeleine Smith; a guilty verdict likely meant death by hanging. There was widespread belief that Smith was certainly guilty of murdering her lover, but the circumstances surrounding the case had also generated enormous sympathy for her. Many asserted that the only tragedy in the case was that Madeleine had to commit the murder herself.

Four months earlier, at about nine o'clock on the evening of March 23, 1857, a young man named L'Angelier left his lodgings in Glasgow, Scotland. Before he left, he paused to talk to his landlady and ask for the passkey, since he didn't plan on being back till very late. The next time his landlady saw him was at two thirty the next morning. L'Angelier did not use the key but urgently banged on the front door and rang the doorbell. The landlady opened the door to find L'Angelier clutching his stomach in pain. He was vomiting violently, and looked so ill that the landlady thought it prudent to call for a doctor. This was not the first time Émile had left the house in good health, only to return a few hours later with an alarming stomach malady. This time, though, would be different. A doctor arrived around seven o'clock and prescribed morphine for the pain. He returned a few hours later to check on his patient, but it was too late. Émile L'Angelier was dead.

Nineteen-year-old Madeleine Hamilton Smith was the daughter and granddaughter of well-known Scottish architects. She was the eldest of five children born into the Smith's upper-middle-class family, and lived at Blythswood Square in Glasgow. Madeleine, a petite dark-haired girl, had been educated in England at Miss Gorton's Academy for Young Ladies, where she learned such important topics as proper manners and the correct way to run a household. Back in Glasgow, Madeleine kept herself busy in the social circuit, attending parties and balls, by some accounts going to five separate parties in one night.

Out walking one day, she met twenty-six-year-old Émile L'Angelier and fell head over heels in love with him. Unfortunately for the couple, the strict mores of Victorian society meant that their relationship was not acceptable.

Pierre Émile L'Angelier passed himself off as French, with all the Gallic charm that he could muster. He boasted of being related to noblemen and royalty who lived in châteaux in the heart of France. In reality L'Angelier was from Jersey in the Channel Islands and not French at all. Far from being of noble birth, he was employed as a lowly clerk in a seed warehouse, earning little more than ten shillings a week. Considering that renting a single room in lodgings typically cost three to six shillings a week, this did not put L'Angelier in the same social class as Madeleine Smith at all.

Despite their different social backgrounds, or maybe because of them, Madeleine adored L'Angelier and started writing him love letters. She sent her first letter to L'Angelier while staying at her family's country residence, and letters continued back and forth between the two once she returned to the city, often arranging to meet "by accident" on the street or at a nearby shop for their trysts. In the heat of their torrid sexual relationship, L'Angelier proposed marriage, and Madeleine readily said yes.

At some point Madeleine's father learned of the relationship and forbade Madeleine ever to see L'Angelier again, instructing her to write a final note to him breaking off the relationship. Not only was L'Angelier penniless, but he was also a "foreigner," totally unsuitable for someone of Madeleine's standing. Madeleine complied with her father's wishes and stopped seeing L'Angelier, but Émile was not put off, and entreated Madeleine to continue seeing him. Surviving letters of Madeleine's show that she was eager to resume the relationship despite her father's prohibitions, writing: "Papa was very angry with me for walking with a gentleman unknown to him. But I don't care for the world's remarks so long as my own heart tells me I am doing nothing wrong." L'Angelier persuaded a female friend to allow him and Madeleine to meet at her house for their romantic interludes. No doubt the forbidden nature of the relationship seemed exciting and attractive to Madeleine, and for two years, the couple continued to meet in secret.

Émile kept all her letters, but he commanded Madeleine to burn all of his, presumably so that they would not be discovered by her father. The few letters of his that do survive reveal him to be a demanding and controlling lover: He dictated what clothes she should wear, where she could go, and with whom she could talk. For her part Madeleine's letters show an insecure young woman desperately seeking approval.

Regardless of these red flags, in 1856 the couple was planning to marry. Not knowing that Madeleine had resumed her illicit relationship, her parents decided that it was time she was married to a suitable husband. Indeed, someone of the proper standing was soon found in William Minnoch. With an annual income of three thousand pounds a year, more than one hundred times L'Angelier's, Minnoch was clearly a much more suitable husband to support Madeleine's lifestyle. Madeleine slowly came to the realization that she was better off wealthy and married to Minnoch than in love but poor with L'Angelier. When Minnoch proposed marriage, she eagerly agreed.

But what about L'Angelier? He still had all of Madeleine's love letters, including those that proved their "criminal intimacy." Madeleine became very worried, realizing the possibility of blackmail should L'Angelier threaten to give the letters to her new fiancé, an action that would destroy her reputation.

Madeleine had been in the habit of talking to L'Angelier through her basement bedroom window, even sneaking him cocoa through the window on chill nights. Sensing an opportunity to solve her problems, Madeleine obtained some arsenic power from a local pharmacist. On Thursday, February 19, as Madeleine and L'Angelier were conversing through her window, she again gave him a cup of hot cocoa to drink. Later, in his lodgings, L'Angelier became very ill; his landlady found him violently vomiting green bile. The next morning, Madeleine left her house and went to Murdoch Brothers Druggists in nearby Sauchiehall Street, where she bought another sixpence' worth of arsenic. That night more cocoa was given to L'Angelier through the window.

Later police would find L'Angelier's diary, in which he had written several damning entries: "Don't feel well"; "Saw Mimi [Madeleine]in Drawing Room . . . Taken very ill"; "I can't think why I was so unwell after getting that coffee and cocoa from her." On the night of March 21 L'Angelier was seen staggering along the street in apparent distress, clutching his stomach and groaning. When he arrived at his lodgings, he was in great pain and vomiting violently. His landlady, Mrs. Jenkins, anxiously sent for the doctor. On the first visit the doctor gave morphine for the pain, but when he returned later that morning, he grimly asked Mrs. Jenkins to close the curtains. L'Angelier was dead.

An autopsy revealed an enormous amount of arsenic in L'Angelier's stomach: almost five grams. At the time no other arsenic murder had shown such huge amounts of arsenic in the victim. It might seem impossible to administer such a large amount of arsenic without the victim noticing, but as pointed out by the prosecution in the subsequent trial, up to six grams (forty times a lethal dose) of fine arsenic powder could be easily mixed with two teaspoons of cocoa and milk or boiling water in a teacup without any untoward smell or taste. Following Madeleine's arrest she was taken to Edinburgh for trial, where she was accused of "wickedly and feloniously" poisoning L'Angelier.

The scandal of premarital intimacy, potential blackmail, murder, foreigners, and fraternization between the classes ensured huge publicity for the trial. The evidence was published almost verbatim in every major newspaper throughout Britain and even in New York. The press was equally divided over Madeleine's guilt or innocence. On the one hand, Madeleine freely admitted to purchasing and using arsenic. She testified that the daughter of a well-known actress had advised her to use arsenic to beautify her complexion; which she did, applying diluted arsenic to her face, neck, and arms.

Of the various theories raised concerning L'Angelier's death, it was suggested that he was an arsenic eater. He was shown to have been familiar with the use of arsenic to improve the stamina of horses, and was known to take arsenic for cosmetic reasons and to improve his breathing. Indeed, one witness at the trial testified that L'Angelier did purchase a powder from a pharmacy on the Sunday night on which he died.

Naturally the defense seized on the notion that L'Angelier was an arsenic eater, invoking the Styrian defense for the first time in a British trial—both victim and accused had valid reasons for their possession of arsenic. After deliberating for nine days, the jury finally reached a verdict of "not proven." In Scottish law this meant that Madeleine hadn't been found innocent, but that the prosecution had not proved her guilt beyond a reasonable doubt.

Throughout the trial three viewpoints were espoused in the press: that Smith was innocent, and her lover had died either by suicide or from an accidental overdose; that Smith had committed murder and should pay the penalty; and, by far the most popular, that Smith probably did it, but that L'Angelier had deserved it.

What was clear after the trial was that Madeleine could no longer live in Scotland. She moved to England with her younger brother, James, changing her name to Lena. There she met and married an artist, George Wardle, with whom she had two children, Tom and Kitten. Lena maintained her wealthy middle-class lifestyle and was a popular hostess. Still courting controversy, she started a fashion for not using tablecloths at dinner, using place mats instead. While today's sensibilities may find this amusingly trivial, it must be remembered that at the time even piano legs were not allowed to be bare, and were covered for modesty. Lena's marriage was not to last, and she and her husband separated after twenty-eight years. When she was seventy, Lena moved to live with her son in New York, where she eventually died at the ripe age of ninety-three.

HOW ARSENIC KILLS

Although most people use the generic term *arsenic* to mean the poison, it is usually not pure arsenic that is involved in murder, but rather compounds containing arsenic. In fact, eating pure arsenic is not likely to do much harm, as it is poorly absorbed by the intestine and rapidly eliminated from the body. Other forms of arsenic are far more deadly.

The introduction of gaslight into Victorian homes made it possible for households to enjoy intense color on their walls. Scheele's green, which was particularly vivid, was all the rage. When this green pigment, made with arsenic, was incorporated into wallpaper, it not only led to brighter homes but also had the added benefit of reducing the incidence of bedbugs. This was a boon to wallpaper manufacturers, who quickly saw the advertising benefits. Unfortunately, what was killing the bedbugs started to affect people too. To stick wallpaper to the walls, a paste of simple flour and water was used. In damp climates the paste was an irresistible food source to mold, especially a mold called *Scopulariopsis brevicaulis*. The mold grew well on the paste and the cellulose that made up the wallpaper, slowly digesting the material as it grew. As the mold metabolized the paper, chemical reactions would convert the solid arsenic in the paper to a volatile arsenic gas called arsine, which has a distinctive garlicky smell. Arsine gas causes red blood cells to disintegrate, reducing the oxygen carried through the body and essentially causing asphyxiation. Knowing that you are unlikely to be bitten by bedbugs as the wallpaper is slowly killing you is not the calm relaxed feeling most people want from their bedrooms. Strangely, though, arsine gas does not cause the same symptoms typically associated with arsenic poisoning.

The toxicity of classical arsenic poisoning is due to the disruption of the cell's normal biochemical reactions. Arsenic compounds are readily absorbed through the intestine, making food and drink the obvious pathway for arsenic poisoning. Arsenic toxicity is primarily associated with two forms of arsenic, arsenates and arsenites, each of which causes death in their own way.

Arsenates chemically and structurally resemble another important molecule, phosphate. In fact, they look so similar that the body is unable to distinguish between them. Phosphates make the backbone of DNA's double helix, they can be stuck onto and removed from enzymes to alter their activity, and as we have seen before, they make up part of the body's energy source of adenosine triphosphate, ATP. Arsenate kills because it gets incorporated into ATP instead of phosphate, but unlike phosphate, arsenate provides no energy for the cell. Think of batteries running a child's toy. A dead battery looks identical to a new one, but a dead battery in a toy will do nothing. In the same way, when arsenate invades a cell, the cell's energy supply of ATP quickly runs down. Without energy to run all the processes and reactions a cell is required to perform, activity eventually stops altogether.

The most common poisonous arsenite compounds are arsenic trioxide, arsine gas, and arsenic pigments, like Scheele's green. Arsenic trioxide is also known as white arsenic, since this is the form that coated the inside of chimneys during ore smelting.

Chemical reactions in the body are carried out by enzymes, and some enzymes are made from amino acids that contain sulfur. Often sulfur amino acids help hold the enzyme together in its proper shape. Arsenites form a strong bond with sulfur, and this prevents the sulfur from doing its job in keeping the enzyme together. As the enzyme falls apart, it can no longer perform its function and stops working. Once arsenites have been ingested and absorbed, they are carried around the body in the bloodstream, potentially interfering with any sulfur-containing enzymes or proteins they encounter.

Since there are a large number of sulfur-containing enzymes

in the body, each with a different function, specific arsenite symptoms can be variable. One of the proteins in the body that contains a lot of sulfur amino acids is keratin, the protein that makes up nails and hair. Often, measuring the amount of arsenic in a hair sample gives a good indication of how much arsenic a body has been exposed to.

Mystery surrounding the death of Napoleon Bonaparte in 1821 has led to several theories concerning his demise. During his last few months in exile on the island of St. Helena, Napoleon was unwell and suffering from extreme stomach pain. Following his death, an autopsy suggested that Napoleon had died from stomach cancer, but it wasn't long before rumors of poisoning were rampant. Naturally the British blamed the French, who blamed the British. In the 1960s, samples of Napoleon's hair, taken from his head shortly after death as souvenirs, were analyzed for their arsenic content. Unusually high levels of arsenic were found in Napoleon's hair, but how had the arsenic gotten there? One suggestion was that it had come from Napoleon's wallpaper. Remarkably, in the 1980s a sample of wallpaper from Napoleon's bedroom was discovered. The wallpaper did contain arsenic in the Scheele green dye, but whether it was enough to kill him is not clear. More likely, as soon as Napoleon felt unwell he called for his doctors, and their recommendations of purgatives and medicines probably did more harm than his wallpaper. Napoleon once famously stated: "You medical people will have more lives to answer for in the other world, than even we generals."

MARSH AND THE DETECTION OF ARSENIC

In the legal cat-and-mouse games of arsenic trials, the accused had a strong refuge in the Styrian defense. However, the prosecution often had a hard time proving death by arsenic; that is, until James Marsh arrived on the scene. Although the 1700s saw the emergence of the fledgling science of analytical chemistry, many arsenic poisoners never went to trial as doctors were more inclined to ascribe death to natural illness rather than to malice. As we have seen, the symptoms of arsenic poisoning and food poisoning are so similar that doctors seldom contemplated poisoning as a cause of death. Chemists had learned to find evidence of arsenic in the organs of cadavers, but the results were often unpredictable and not always reproducible; not something on which many lawyers were ready to stake their cases in court. In 1832 John Bodle was charged with murdering his eighty-year-old grandfather, George Bodle. A maid who worked at the grandfather's farm testified that John had told her he wanted George dead so that he could get his hands on the estate, worth £20,000 (or £2,300,000 today).

A local pharmacist confirmed that John had indeed purchased a quantity of arsenic in the days before George's demise. James Marsh, a young chemist, had been asked to testify at trial on behalf of the prosecution, showing that arsenic had been found in some suspect coffee that John had given his grandfather, as well as in several organs removed postmortem. At the time the standard test for arsenic was to bubble hydrogen sulfide gas through a solution of arsenic to make arsenic sulfide. Indeed, Marsh did obtain a precipitate of arsenic sulfide, proving that arsenic was in George Bodle's tissues, but by the time the trial came around the precipitate was so discolored that the defense convinced the jury that the evidence was worthless. Most court cases then were based on the character of the accused. Young John seemed a likable man, certainly much nicer than his father, who also had access to the arsenic rat poison. With all the character evidence, the judge handed down a verdict of not guilty, and young John was set free.

In a final mocking of the forensic evidence, Bodle did eventually confess to murdering his grandfather, sure in the knowledge that he could not be tried twice for the same crime. Marsh was frustrated that he was unable to provide convincing evidence of Bodle's guilt. Returning to his laboratory, Marsh worked relentlessly for several years until he arrived at a foolproof way of detecting arsenic in bodies, which he did successfully in 1836. Marsh's method began by finely mincing the victim's tissues and heating them with strong acid to destroy the organic matter and bring any arsenic into solution. The next step was to convert the arsenic into a gas by adding small amounts of zinc to the acid solution, generating arsine gas. Arsine was then broken back down to arsenic and hydrogen through heating, and any arsenic present would then condense and collect on a porcelain or glass plate as a gray metallic film. The amount of arsenic present in the sample could be determined by weighing the glass plate before and after exposure to the test tissue.

ARSENIC AS THERAPY

Intriguingly, the ancient history of arsenic as a tool of murder is matched only by its ancient history as a tool of healing. Hippocrates (460–377 BC), one of the best-known figures in the history of medicine, used the mineral realgar, a ruby-red crystalline rock of arsenic and sulfur, as a remedy for ulcers on the body.

In June 1771 Thomas Wilson of London was granted a patent for a remedy entitled "Tasteless Ague and Fever Drops." Ague (likely malaria or some other illness involving fever and shivering) was endemic in many parts of England. Certain parasitic organisms, such as those responsible for malaria, do seem to be particularly sensitive to low doses of arsenic, lower than what would typically harm a person over the short time of treatment. Indeed, arsenic solutions were also the first treatments available for another particularly nasty parasite, the one causing syphilis.

Not one to be shy in marketing his remedy, Wilson promoted his patent medicine as "a medicinal composition which, after much experience, hath been found to be an infallible remedy for agues and intermittent fevers, even in the most obstinate cases where the bark [e.g. aspirin from willow], and every other medicine hath proven ineffectual." Nonetheless, Wilson's "Fever Drops" appeared to work and were adopted by hospitals across England.

Thomas Fowler, a physician at the Infirmary of the County of Stafford, in the English Midlands, was so impressed with the effectiveness of the Fever Drops that he persuaded the apothecary at the infirmary to analyze the drops and find out what was in them. It was discovered that the active ingredient was none other than arsenic. Dr. Fowler devised his own version of the medicine, and with easy access to patients on whom to try his concoctions, Fowler eventually had enough findings to put together a book. In *Medical Reports of the Effect of Arsenic in the Cure of Agues, Remitting Fevers, and Periodic Headaches*, Fowler reports that in 271 cases of ague, 171 patients were "cured" by his solution. An aggressive marketer, Fowler realized that having his remedy associated with arsenic, which everyone knew was a poison, was probably not the best sales pitch, and so introduced his medicine to the public as "The Mineral Solution."

Fowler's original solution contained arsenic trioxide, distilled water, and vegetable extract. Lavender oil was also added to give it a more "medicinal" appearance. The mineral solution quickly became known as Fowler's Solution and was rapidly adopted into medical practice for the treatment of epilepsy, hysteria, melancholy, dropsy (edema), syphilis, ulcers, cancers, and dyspepsia (indigestion). And of course nothing promotes a product better than celebrity endorsements. James Begbie, vice president of the Royal College of Physicians in Edinburgh, and physician to Queen Victoria when she was in Scotland, wholeheartedly touted the health benefits of Fowler's Solution, almost guaranteeing its widespread popularity.

While Fowler's Mineral Solution likely was effective against things like syphilis and certain cancers, the problem with many eighteenth- and nineteenth-century medications was the erroneous assumption that just because a drug worked well on one disease, it was a panacea. It is highly unlikely that the solution would provide any benefit for symptoms as vague as melancholy and hysteria. However, its widespread use in eradicating vermin during the eighteenth and nineteenth centuries was nonetheless beneficial, as it helped reduce diseases carried by fleas living on mice and rats. And despite arsenic solutions falling out of favor as a therapeutic option in modern medicine, they have recently seen a resurgence in interest for the treatment of some leukemias.

Certainly arsenic was a start on the path toward improving public health, but the next chapter describes a chemical whose impact on reducing disease in growing towns and cities cannot be overstated. Indeed, it is likely that this chemical can be found under most kitchen sinks.

. { 11 } .

Chlorine and the Killer Nurse of Lufkin

In peace-time, the scientist belongs to humanity;
in war-time, to his fatherland.

—Fritz Haber, Nobel Prize–winning chemist, 1918

CHEMICAL WARFARE

The early twentieth century saw huge social and political changes throughout Europe. Queen Victoria died in 1901, ending the longest reign in the British monarchy until that of the present Queen Elizabeth II. The year 1914 also saw the start of the largest European war yet. It was during this "war to end all wars" that one of Germany's top scientists created the first weapon of chemical warfare.

Fritz Haber was born in 1868 in Breslau (now Wrocław, Poland). He studied chemistry in Berlin, hoping to transform himself from a provincial Jewish boy into a successful German. At the outbreak of World War I, Haber, working as director of the Kaiser Wilhelm Institute of Physical Chemistry in Berlin, was desperate to prove his patriotism, and willingly became a uniformed consultant to the German War Office.

Haber was convinced that chemistry could be used to drive the Allies from their trenches and allow a German victory; the tool he would use to empty the trenches was poisonous gas. He decided that chlorine gas would make a most effective weapon, but the problem was how to deliver it. One early test of chlorine dispersal had resulted in the deaths of several German troops. German army officers were not as convinced as Haber regarding the use of chemical weapons, with one general calling them "unchivalrous," and another declaring that "poisoning the enemy as one poisons rats is repulsive." But by 1915 defeats on several fronts hardened the German army's resolve to deploy chemical weapons.[1]

After waiting weeks for ideal wind conditions—strong enough to carry the chlorine gas away from the German trenches into the Allies', but not strong enough to disperse the gas—Haber released some 168 tons of chlorine onto Allied troops in trenches in Ypres, Belgium. A sickly cloud "like a yellow low wall" that smelled like a mixture of pineapple and pepper drifted toward the Allied trenches.

At first the green clouds were thought to be a smokescreen from behind which the German infantry would attack, but when the gas arrived at the trenches, no one was prepared for what would happen. Being heavier than air, the chlorine naturally flowed and sank down into the trenches. As the soldiers breathed in the gas, they complained of chest pains and burning sensations in their throats. The attacks were later described as "an equivalent death to drowning, only on dry land. The effects are there—a splitting headache and terrible thirst (to drink water is instant death), a knife edge of pain in the lungs and the coughing up of a greenish froth from the stomach and the lungs, ending finally in insensibility and death. It is a fiendish death to die." Some ten thousand troops were affected by the gas, almost half of whom died from asphyxiation within ten minutes of the chlorine entering the trenches.

Haber was giddy with delight over his chemical weapon, even developing what he self-aggrandizingly called Haber's Rule, a mathematical model describing the relationship between gas concentration, exposure time, and death rate. Although traditional weapons killed far more people in World War I than did chlorine gas, the new chemical weapons added a dreadful new dimension to the horror of war.

WHY CHLORINE IS TOXIC

Although it is unlikely nowadays that anyone will be deliberately exposed to chlorine gas, anyone who has ever been swimming in an over-chlorinated pool knows how irritating it can be to the skin and eyes.

A thin layer of fluid covers the tissues of the eyes, nose, mouth, and lungs. This thin layer is critical to keeping the organs moist and working properly. Our tears help prevent inflammation, infection, and scarring. Films of saliva in the mouth also contain mucus and antibiotics to lubricate food as we swallow, and also to kill bacteria that can lead to ulcers and cavities. In the nose and airways, the thin liquid layer is particularly sticky, trapping dust, viruses, and bacteria that might otherwise get into the lungs and cause infections. Of course these defense mechanisms can become overwhelmed when exposed to large amounts of bacteria and viruses, but normally they work pretty well. However, it is this thin layer of liquid, designed to protect us, in which chlorine gas dissolves to cause problems.

When chlorine dissolves in water, it forms two acids: hypochlorous acid and hydrochloric acid. Our bodies are very familiar with hydrochloric acid, as this is the acid made in the stomach to help kill any bacteria we ingest and to start the breakdown of food. Although the stomach produces concentrated hydrochloric acid, it also takes great lengths to protect itself from the acid it makes. A thick layer of mucus is spread over the lining of the stomach, providing a physical barrier between acid and the cells lining the stomach.

These inbuilt precautions do not exist in the eyes or the lungs, and so chlorine and its acids have direct access to these tissues. When dissolved in the thin liquid film covering the eyes, the chlorine acids attack, causing painful irritation and even temporary blindness. However, the eyes do have one protective mechanism, and that is tear production. When the eye is irritated, a flow of tears is started to wash away the irritant. If a person's eyes are not exposed to chlorine for too long, the tears will eventually wash away the hypochlorous and hydrochloric acids, allowing the eventual recovery of the eyes and sight.

The lungs, on the other hand, have little protection. Acids

formed by chlorine inhalation cause severe irritation and damage to lung tissue. The immediate effect is constriction of the airways to reduce the noxious chlorine getting farther down into the deep recesses of the lungs, where carbon dioxide and oxygen are exchanged. Unfortunately this also restricts the flow of oxygen, causing great difficulty in breathing, as the victim gasps for air—only to bring more chlorine into the lungs.

Irritation of the lung tissue also starts the coughing reflex. This is normally a good thing, as it helps get rid of small bits of debris or bacteria by forcefully expelling air from the lungs. But with chlorine exposure, this normal coughing response goes into overdrive, causing severe and prolonged coughing fits, which also make breathing difficult. As the delicate cells lining the airway and lungs start to die off, inflammation can quickly damage delicate lung tissue. Many soldiers who escaped death by chlorine and survived the war had breathing difficulties for the rest of their lives. With inhalation of high amounts of chlorine, the damage to the lungs is so severe that excess fluid from the blood vessels surrounding the lungs slowly accumulates inside. Death from asphyxiation follows, as the victims literally drown in their own fluids.

There is no antidote for chlorine poisoning. Removing the victim from chlorine exposure is the most immediately important step to take. After that, ensuring that patients continue to breathe is the only thing that can be done. Death can be relatively quick or agonizingly slow, depending on how much chlorine the person has been exposed to, and the extent of the damage sustained.

CHLORINE TO THE RESCUE

While the hypochlorous acid produced by chlorine can cause severe damage to the human body when used inappropriately, it has also been one of the greatest boons to public health.

In nineteenth-century Paris, there was a huge demand for animal intestines, which were used to produce strings for musical instruments and for goldbeater's skins (sheets of intestines used in the formation of gold leaf). Intestines were prepared in "gut factories," which were very smelly places. They were also quite dangerous, given the large quantities of germs released from animal intestines. The problem was so bad that in 1820 the French Society for the Encouragement of National Industry offered a reward to anyone who could come up with a means to process the intestines without the putrefaction.

The prize was awarded to Antoine Germain Labarraque, who discovered that bubbling chlorine through water made a solution—hypochlorous acid—that would prevent the stench of decay, and even prevent the decomposition in the first place. Labarraque's solution went on to be used in latrines, sewers, slaughterhouses, anatomy laboratories, prisons, and morgues. Labarraque also recommended that doctors wash their hands in chlorinated lime, and also sprinkle it on patients' beds in cases of contagious infection. Unfortunately, he also recommended that doctors inhale chlorine before seeing their patients.

Probably the most publicized use of Labarraque's chlorine solution was in 1847, when Dr. Ignaz Semmelweis used chlorine to "deodorize" the hands of Austrian physicians. Semmelweis had noticed that doctors "carried the stench of decomposition from the autopsy room to the delivery room," noting that death rates during delivery in hospital by doctors were significantly higher than among women with midwives, or even those delivering on the street.[2] Although Semmelweis was initially ridiculed for his views, his use of Labarraque's solution started the now-common practice of hand washing to stop the spread of disease. Labarraque's chlorine solution is now used throughout the world, wiped on countertops and sinks and thrown in with the laundry. It is more commonly known as household bleach.

DEATH BY BLEACH

In the United States around 15 percent of adults suffer from chronic kidney disease. If left untreated, this can lead to stroke, heart attack, or even death. For those affected, dialysis—essentially an artificial kidney that cleanses the blood for those whose kidneys can no longer perform the job—is a literal lifesaver. One of the big players in the dialysis business is the DaVita corporation, a Denver-based company whose name means "giving life" in Italian. Certainly DaVita is a lifeline for those in need of medical care. In early 2008, however, one nurse not only tapped into that lifeline but callously exploited it, using as a murder weapon the very equipment meant to give life.

Kimberly Clark Saenz was born in 1973 in Texas to a blue-collar family; her father worked for a trucking company, and her mother was employed at the local Walmart. Following a hospitalization for pneumonia, Kimberly decided that her future career path would involve taking care of other people, the same way she had been cared for in the hospital. She enrolled in a local community college, graduating as a licensed vocational nurse (LVN).

Even from the start, Kimberly was not destined to win any employee-of-the-month awards. Within a two-year phase, she had lost jobs at two hospitals, an assisted living center, a doctor's office, and a home health care facility. During her employment at the Woodland Heights Medical Center in Lufkin, 120 miles northeast of Houston, staff and administrators noticed that controlled drugs were disappearing, and investigations finally led to Saenz. Her purse was found to be full of Demerol, an opioid narcotic. Not only was Saenz stealing drugs, but she was faking her own urine tests to cover her drug abuse. Naturally Saenz was asked to leave her job at the hospital, and the Texas Board of Nursing began an investigation into Kimberly's actions.

Because the board was still investigating Saenz, their deliberations were not open to view by potential employers. When she was hired by a DaVita dialysis clinic, neither her employer nor her patients had any idea of the trail of problems she had left behind. Although Saenz was only licensed to give medications, she was often needed as a patient-care technician, connecting the patients to the dialysis machines and attending to their needs during the procedure. Saenz apparently felt that simply hooking up patients to the dialysis machines was beneath her. Not only did she complain to other staff members about her poor treatment by her employer, she also expressed her antipathy toward some of the patients under her care. One employee testified that Saenz had taken a particular dislike to five patients. Coincidentally, all those patients either died or were injured while undergoing treatment at Saenz's hands.

The kidneys do many things for the body, helping to keep it healthy by regulating the internal environment. The kidneys play a major role in controlling blood pressure, and making calcitriol, the active form of vitamin D needed to absorb calcium from the diet. Kidneys also make the hormone erythropoietin, or EPO, which triggers the production of red blood cells. Roughly twenty times each day, the kidneys filter the body's entire blood supply. As the blood is filtered, things that the body wants to keep, like sugar and amino acids, are reabsorbed into the blood. Impurities in the blood that need to be removed are sent to the bladder for excretion through urination. For people with kidney failure, dialysis machines essentially take over the job of the kidneys, and patients go every other day or so to have their blood filtered at a dialysis center.

The DaVita dialysis center in Lufkin, Texas, like most medical facilities, appeared bright and clean, with a faint but instantly recognizable antiseptic odor. Bleach is the main disinfectant and sterilizer used at DaVita, and each week, staff members run a bleach solution through the dialysis machines to remove any harmful bacteria that may be lurking within. After bleaching, the machines are thoroughly rinsed with large amounts of water, to remove any remaining bleach. Bleach is used on the floor to clean occasional drops of blood, and every time a patient finishes a treatment, the chair, machine, and surrounding area are wiped down with a bleach solution.

Dialysis is not a quick procedure. Most patients spend three or four hours every other day doing treatments. During treatment, DaVita requires patients' vitals to be checked every thirty minutes.

Clara Strange was scheduled for dialysis at the Lufkin DaVita center on Tuesday, April 1, 2008. At 11:34 a.m., she was hooked up to the dialysis machine, which started cleaning her blood. Throughout the rest of the morning and into the early afternoon, Ms. Strange seemed to be doing well. When her patient-care technician returned from a scheduled thirty-minute break, he was horrified to find Ms. Strange slumped in her chair, unresponsive and without a pulse. The technician yelled for help, and the crash cart was rapidly brought to Ms. Strange's side. Nurses and doctors converged on the patient, but despite their attempts to revive her, she was dead from cardiac arrest, still attached to the dialysis machine.

Thelma Metcalf was slated for her dialysis appointment at the same time as Clara Strange, and the two shared the same dialysis bay. She arrived in a good mood, talkative and happy to see her friends at the center. At 3:05 p.m., just minutes after Clara Strange was declared dead, Metcalf was also found unresponsive and without a heartbeat. The crash cart brought in to try to resuscitate Strange was still in the bay, since no one had had time to put it away. As the medical team tried to get Metcalf's heart beating again, Kimberly Saenz was repeatedly asked to help with Metcalf's breathing, but she appeared distant and uninterested.

The EMTs arrived and rushed Metcalf to the hospital. In the ambulance, three rounds of adrenaline were pumped into Metcalf to try to get her heart started again, but it was too late. Metcalf was dead on arrival from cardiac arrest.

The chances of a patient suffering cardiac arrest during dialysis are seven per hundred thousand dialysis sessions. The odds of it happening by chance to two patients within minutes of each other, would be more than one in a billion. The chances of winning the lottery are much better—only one in 300 million.

Inspectors from the state and DaVita's corporate office came in to investigate. Lack of proper personnel training, poor record keeping, and inconsistent sterilization were noted, but no one suggested foul play.

On April 16 fifty-nine-year-old Garlin Kelley arrived to start his dialysis. Kelley was an early bird, and by 5:36 a.m. he was hooked up to the dialysis machine and doing fine. Two hours later Kelley was still well. At 7:35 a.m. Kelley's med nurse, Sharon Dearmon, had her back to Kelley, tending to another patient. Suddenly the alarm on Kelley's dialysis machine sounded, blaring its warning into the room. Dearmon whipped around to see Kimberly Saenz frantically trying to reset the alarm. Dearmon rushed over to see what was going on, and found Kelley slumped and unresponsive in his chair. Dearmon yelled for help and shut off Kelley's dialysis lines, then started CPR.

One of the people to respond to Dearmon's call for help was RN Sharon Smith. When interviewed later, Smith recalled noting something strange in the bloodline of the dialysis machine. It looked like an unusual blood clot, but it was fibrous, almost like hair. "I've never seen it before, and I've never seen it since," said Smith. Dearmon also recalled the odd-looking brownish clot. Still unconscious, Kelley was transported to the hospital, where he remained in a coma for four months, dying without ever waking.

The Lufkin DaVita dialysis center had instituted a policy requiring the collection and storage of all the IV lines and syringes used on patients who had experienced cardiac complications during dialysis. Generally, this had never been followed until the deaths of Thelma Metcalf and Clara Strange. On April 16 the policy was followed to the letter, and Garlin Kelley's bloodline, with a syringe still attached, was placed in a bag and put in the freezer. Later forensic examination would reveal the unmistakable presence of bleach.

On April 28, 2008, Marva Rhone entered the dialysis clinic, and by 5:52 a.m. was hooked to the dialysis machine, ready for the next few hours of treatment. By 8:15 things were rapidly going downhill. Rhone's blood pressure was dropping, and she appeared to be very uncomfortable, squirming in her chair and suddenly vomiting. Rhone tried to speak, but her voice was weak, and her speech was becoming slurred. Remarkably, the clinic staff was able to stabilize Ms. Rhone. Subsequent blood work at the hospital revealed high levels of potassium and an enzyme called lactate dehydrogenase, or LDH. Taken together, these are suggestive of massive damage to cells in the body.

Although the cause of Strange, Metcalf, and Kelley's sudden deterioration while undergoing dialysis remained unclear, the cause of Ms. Rhone's decline had been observed as it occurred.

Another patient scheduled for dialysis on April 28 was

sixty-two-year-old Lurlene Hamilton, an eight-year veteran of dialysis treatments. For three years, Lurlene had been a regular outpatient to the Lufkin facility for her dialysis treatments, and was familiar with the normal routines. As she was undergoing dialysis, Hamilton had observed Nurse Kimberly Saenz approach Marva Rhone's dialysis machine. That was not unusual in itself, but the way in which Saenz ambled over to Rhone was notable: Saenz appeared to be looking around to see if anyone was watching her.

What happened next was totally unexpected to Hamilton, who watched as Saenz poured bleach into a bucket on the floor. She had no doubt that it was bleach, as the unmistakable acrid fumes made their way across the room to her dialysis bay. She watched in horror as Saenz calmly filled syringes with bleach and then injected the contents into Rhone's dialysis line.

Hamilton was not the only person to witness this macabre scene. Next to her was Linda Hall, another dialysis patient. Hall, too, saw Saenz fill syringes with bleach, then calmly inject them into Rhone's dialysis port. Hall couldn't believe what she was seeing. Had she really witnessed a nurse injecting bleach into a patient's IV line? More terrifying was the fact that Kimberly Saenz had been assigned to look after *her*! Both Hamilton and Hall waved frantically to get the attention of another technician, begging her not to let "Kim" touch them. The technician had two terrified patients—and no idea what to do next.

The obvious step was to report up the chain of command, and that person was Registered Nurse and Clinical Coordinator Amy Clinton. When Clinton was told what the two patients had witnessed, she couldn't believe it. Clinton confronted Saenz, who denied giving anyone medication or injecting bleach into IV lines. After Saenz was sent home for the day, Clinton examined the bucket and syringes that Saenz had used that day: All tested positive for bleach.[3]

On April 29 Saenz was fired from the DaVita Lufkin Dialysis Center, and the center itself was shut down for two months. DaVita released a statement saying: "We believe the events that led to our voluntarily closing the Lufkin Dialysis center were the result of a criminal act by an individual who has been terminated and is no longer working at the center." On May 30, 2008, the Lufkin police arrested Kimberly Saenz. Forensic examination of Saenz's home computer showed a Yahoo search for "bleach poisoning," which had directed her to an article on chlorine poisoning. Earlier searches for "bleach given during dialysis" and "can bleach be detected in dialysis lines" were also found.

In fact bleach was detected in all the syringes and dialysis lines that had been collected from Saenz's deceased patients. Altogether, Saenz was charged with five counts of aggravated assault on five separate individuals—Marva Rhone, Carolyn Risinger, Debra Oates, Graciela Castañeda, and Marie Bradley. In each case Saenz was charged with introducing bleach into the victim's bloodstream. In a sixth count, Saenz was charged with the capital murders of Clara Strange, Thelma Metcalf, Garlin Kelley, Cora Bryant, and Opal Few, again by injecting bleach into their blood.

Kimberly Saenz's trial lasted seventeen days and included testimony from forty-nine witnesses and the introduction of almost four hundred exhibits. Garlin Kelley was one of only two victims to survive long enough to be taken to the hospital for treatment. As a result, Kelley's blood had been tested for 3-chlorotyrosine, a chemical made only when bleach interacts with tyrosine, an amino acid found in most proteins in the body, including the blood protein hemoglobin.[4] The test for chlorotyrosine was positive.

One expert testified that the levels were higher than anything he had ever seen, and three to four hundred times greater than would be expected to be found in a person undergoing dialysis. A physician and toxicologist with the Centers for Disease Control concluded that there was clear evidence of bleach in the syringes and IV lines, and that the victims had died as a direct result of bleach injections. When questioned whether he could say how much bleach was in the victim's blood, he replied that it was impossible to measure because bleach reacts too quickly, converting to hypochlorous acid, which then damaged the victim's organs and tissues.

The jury found Saenz guilty of aggravated assault on three counts, and guilty of capital murder on five. She was sentenced to twenty-three years for each count of aggravated assault, and life in prison without the possibility of parole for the capital murder charges.

INTRAVENOUS BLEACH

We have seen that chlorine gas, when dissolved in the thin fluid layers of the body and forming hypochlorous acid, is certainly noxious. But what if that same acid, in the form of bleach, were directly injected into the blood?

When bleach is injected into the bloodstream, it encounters red blood cells. There are roughly two and a half trillion red blood cells in a single pint of blood. Anything that damages these cells will greatly impact the blood, and the rest of the body. The protective membrane surrounding the red cells is obliterated by bleach, in a process called hemolysis (literally, "breaking of red cells"). As the proteins from the cells are released into the blood and encounter more bleach, they become unraveled. These unfolded protein strands intertwine into long strands like thousands of sets of Christmas lights all tangled together. The iron in the hemoglobin, which gives blood its characteristic color, is exposed, giving the strands a brownish, rusty tinge. Solid aggregates of tangled blood proteins can block small arteries and veins, and may even block the arteries that supply the heart with blood, causing a heart attack. The formation of such clots were likely seen by Nurse Sharon Smith as she later recounted the strange material she spotted in Garlin Kelley's dialysis line.

Bleach also causes dangerous chemical reactions within the blood. When bleach encounters blood proteins, it forms the chemical formaldehyde. This is the chemical that cadavers are saturated with to preserve them for medical school dissections. Formaldehyde easily gets into cells, and there cause all the proteins in the cell to link up with one another, forming a rigid mesh and killing the cell instantly.

While all this is happening in the blood, the circulation is quickly bringing the injected bleach up to the heart. Since red cells are a major reservoir for body potassium, destruction of large numbers of these cells can quickly increase potassium in the blood. We saw previously what disastrous effects large amounts of potassium in the blood can have on the heart.

LETHAL LEMONADE

The bitter taste of alkaloids can be masked, the slight almond aroma of cyanide may not be perceived by everyone, and arsenic really has no taste. But the chlorine smell of bleach is unmistakable, so few people would unknowingly drink bleach. However, this has not deterred the use of bleach in attempted poisoning efforts by less-than-clever criminals.

In July 2010 nineteen-year-old Larenzo Morgan of Caruthersville, Missouri, was angry with his girlfriend. With twisted logic, Morgan decided that the best way to get back at his girlfriend was by poisoning her and her young children. Morgan poured bleach into a pitcher of lemonade sitting in the fridge and filled the ice tray with water and bleach. Later, when two of his girlfriend's children came in for a drink after playing outside, they poured a glass of lemonade each. One sip, and each spat out the disgusting beverage.

Clearly the adulteration of the lemonade was deliberate, but the children hadn't swallowed much, and there seemed little point in going to the ER, or calling the police. In fact the whole episode might have gone unreported had the father of another child living in the house not decided to go after Morgan. That's when the police got involved. Morgan admitted to attempting to poison the children, and was charged and ultimately convicted of first-degree endangerment of the welfare of a child.

Chlorine bleach is yet one more example of a chemical that has been of great benefit in preventing death and disease when used appropriately but that can kill when used improperly. The fact that bleach can be found on grocery store shelves neither diminishes its toxic nature nor reduces the care required when using it.

Epilogue: The Garden of Death

Tobacco, coffee, alcohol, hashish, prussic acid, strychnine,
are weak dilutions: the surest poison is time.
—Ralph Waldo Emerson, "Old Age,"
Atlantic Monthly, January 1862

Nestled in the rolling Northumberland hills of northeastern England is Alnwick Castle. The castle, used as a backdrop in several Harry Potter films, possesses an unusual attraction that would not be out of place in Harry's Hogwarts. Among the manicured formal gardens and cascading fountains is a garden surrounded by high walls and guarded by ornate heavy iron gates. Emblazoned above the entrance is a sign warning visitors, "These Plants Can Kill." Escorted visitors are prohibited from smelling, touching, or tasting any of the plants lest they fall victim to the deadly greenery. The garden now boasts more than one hundred different species of deadly plants, including cannabis and cocaine-producing plants, and is designed as part of the current duchess's educational mission on the dangers of drugs. Many of the plants described in this book, including belladonna, brugmansia (a plant related to jimsonweed), aconite, and castor plants, are given prominence in this garden of death.

In ancient times, and even through the early eighteenth century, murder by poison was fairly easy to get away with. Many of the symptoms associated with poisons were remarkably like those of prevailing infectious diseases, particularly the ones that caused gastrointestinal distress. In such cases murders were often attributed to death from natural causes. Even when a death was suspicious, there was no way to detect the poison and prove that a murder had been committed.

It was not until the eighteenth century that enough scientific progress was made for methods of detection to become elucidated. But even when poisons could be revealed in a test tube, identifying the same compounds in a corpse added an extra level of difficulty that was not always easy to overcome. Nevertheless, it is now possible to determine not only if a poison is present in a victim's body, but also how much. While murder by poison is much rarer than it used to be, this book shows that it still occurs. However, the likelihood of getting away with such murders in the present day is almost nonexistent. Despite the toxic, and even lethal, effects of chemicals found in plants, the chemicals themselves are not intrinsically good or bad. It is only the uses to which they are put that makes them beneficial or deadly. Curiously, much of our modern understanding of how the human body works has come about by using poisons. Understanding the electrical signaling in the heart, for example, has been greatly helped by the use of digoxin and related chemicals, paving the way for the development of better and more specific drugs to treat arrhythmias and cardiac arrest.

The application of poisons like atropine and related compounds like nicotine (the same toxic chemical found in cigarettes) to human tissues has provided insight into how signaling in nerves works. It will be remembered that one of the symptoms of atropine poisoning is an excessively dry mouth. We now know that this is because atropine interferes with the signals controlling salivation and the release of fluid into the airways to keep them moist and functioning properly. Patients who are unconscious or undergoing surgeries, particularly when they have to be intubated, run the risk of excess saliva falling down the back of the mouth and down into the lungs. Too much fluid in the lungs not only makes breathing difficult but can lead to infections like pneumonia. Doctors will use atropine on intubated patients to help dry up saliva and prevent life-threatening lung infections from occurring.

Learning how one deadly substance gets inside a cell can often provide clues as to how other things also get into the cell. For example, ricin gets into cells using a process referred to as endocytosis. The same process used by ricin is also used by viruses in the intestine, like rotovirus, and viruses that infect the airways, including coronavirus. Using poisons like ricin to figure out how chemicals get into cells is the first step in understanding not only how that entry can be prevented but also to improve how drugs can be taken up more efficiently, so patients can gain the same benefit with less medication.

Useful poisons are not restricted to chemicals from plants. Potassium can cause sudden cardiac arrest, but research into how potassium affects the electrical currents across cells has also been crucial in determining how insulin is released from the pancreas. Drugs called sulfonylureas are taken by millions of people with type 2 diabetes to help them release more insulin into their blood. These drugs work by altering how potassium moves out of the cell, and their creation stems from research in which high levels of potassium were applied to pancreatic cells. Overall, the use of poisons in scientific research is probably more widespread than most people realize. Without poisons our understanding of how the human body works would have been greatly hampered.

While some scientists have dedicated their careers to detecting poisons, it is true that others have decided to use their skills to create novel poisons for their own nefarious purposes. Yet it is often the very exotic character of those poisons that allows them and their killers to be identified. Even so, the drugs themselves harbor no intrinsic malice: They are merely chemicals. Though it is disturbing that so many scientists and health care providers have used their knowledge not to help people but to harm them, it cannot be overstated that the responsibility for such atrocities lies not with the drugs, but solely with the killers who use them.

Appendix: Pick Your Poison

Author's Note: The following information is purely for educational purposes only, and is not intended to give the advantages or disadvantages for the use of any particular poison in the commission of murder.

Note: 1 tsp is around 5,000 mg

ACONITE

Route of entry: Ingestion

Lethal Dose: Around 2 mg

Target: Alters signaling along the nervous system

Symptoms: Nausea, vomiting, diarrhea, burning, tingling, and numbness in the mouth and face, spreading to the limbs; sweating, dizziness, respiratory distress, delirium, paralysis of the lungs and heart

Antidote: None specific; may use cardiac drugs to counteract effects

ARSENIC

Route of entry: Ingestion

Lethal Dose: 40–100 mg

Target: Potentially every sulfur-containing enzyme in every cell of the body; halts energy production and cellular repair

Symptoms: Violent vomiting and diarrhea, abdominal pain, muscle spasms, difficulty swallowing, intense thirst, soreness in mouth and throat with difficulty swallowing, weak pulse, kidney failure, coma; death within 12–36 hours

Antidote: Dimercaprol, also known as British anti-Lewisite, binds tightly to arsenic and renders it inactive; can also be used to treat acute poisoning by mercury, gold, and lead

ATROPINE

Route of entry: Usually ingestion

Lethal Dose: In excess of 50 mg

Target: Neurotoxin, blocks normal synaptic transmission by blocking acetylcholine receptor

Symptoms: Extreme dryness of the mouth, slurred speech, hallucinations, blurred vision, sensitivity to light, delirium, urinary retention, rapid heart rate, respiratory paralysis

Antidote: None specific, though physostigmine may be used to counteract some effects

CHLORINE

Route of entry: Injection and inhalation

Lethal Dose: 34–51 parts per million in the air as chlorine gas; 20 g for an oral dose; 2 g for an IV dose

Target: Blood cells, muscle, delicate tissues of airways, nose, eyes

Symptoms: Injection leads to breakdown of blood cells, causing anemia, reduced oxygen delivery to kidneys and brain; oxidation damage to blood proteins; inhalation leads to chemical burns of throat, airways, and lungs, leading to respiratory distress; fluid collects around the lungs, making breathing difficult

Antidote: None

CYANIDE

Route of entry: Inhalation and ingestion

Lethal Dose: Around 500 mg

Target: Targets the mitochondria, shutting off energy production

Symptoms: Convulsions, low blood pressure, low heart rate, coma, lung damage, respiratory failure, cardiac arrest

Antidote: Cobalt salts such as dicobalt-edetateor vitamin B12

Note: Cyanide is one of the fastest-acting poisons known to man

DIGOXIN

Route of entry: Ingestion or injection

Lethal Dose: Few mg

Target: Causes blockage of electrical signaling in the heart

Symptoms: Dizziness, confusion, hallucinations, abdominal pain, muscle pain, weakness, nausea, vision changes, irregular heartbeat, palpitations, breathing difficulties, cardiac arrest

Antidote: Atropine or Digibind (antibodies that mop up excess digoxin)

INSULIN

Route of entry: Injection only

Lethal Dose: 400–600 units, equivalent to 13–31 mg

Target: Liver, muscles, adipose insulin receptors, causing drastic drop in blood sugar

Symptoms: Sweating, vomiting, weakness, irritability, confusion, coma

Antidote: IV glucose

POLONIUM-210

Route of entry: Ingestion

Lethal Dose: Around 0.0005 mg

Target: DNA in the nucleus of each cell

Symptoms: Severe headache, diarrhea, vomiting, hair loss, widespread damage of all internal organs; death within days to weeks

Antidote: None

Note: Around a million times more deadly than cyanide

POTASSIUM

Route of entry: Ingestion or injection

Lethal Dose: 2,000 mg by injection; oral potassium is less toxic and needs around 400,000 mg

Target: All cells, but cardiac cells are particularly vulnerable

Symptoms: Nausea, vomiting, lethargy, numbness, chest pain, difficulty breathing, irregular heartbeat, cardiac arrest
Antidote: None, but treatment includes dialysis and diuretics to help the kidneys excrete excess potassium

RICIN

Route of entry: Injection, inhalation, or ingestion
Lethal Dose: Around 1.5 mg
Target: Protein synthesis machinery in every cell
Symptoms: Injection leads to fever, nausea, hemorrhaging, widespread tissue damage, and organ failure; inhalation symptoms appear 4–8 hours after exposure with inflammation and bleeding in the airways and lungs; fever, cough, tightness in the chest, leading to weakness and fluid buildup and finally respiratory failure; ingestion causes nausea, vomiting, bloody diarrhea, intestinal bleeding, and shock; death within 3–5 days
Antidote: None
Note: Ricin is one of the most toxic substances known to man

STRYCHNINE

Route of entry: Injection, ingestion, or absorption through eyes and mouth
Lethal Dose: 100–140 mg, or 2/100 tsp
Target: Neurotoxin, targeting glycine receptors
Symptoms: Violent convulsions, asphyxiation, hyperthermia (from muscle contraction), tetanic spasms leading to

opisthotonos (see pages 86–88); if the victim survives the initial effects, muscle breakdown can cause kidney damage, along with permanent nerve damage

Antidote: None specific

Note: One of the most painful ways to die; victim dies from exhaustion and asphyxiation 3–4 hours after exposure

Acknowledgments

A huge thank-you to my wife and daughters for their continued support and encouragement during the writing of this book. You are a constant source of joy and happiness. I hope my wife will finally be convinced that the many scribbled notes on poisons that she found lying around really were for a book! I am also indebted to my parents for their lifelong support, particularly during my undergraduate and graduate studies in biochemistry, even when they were not quite sure what biochemistry was.

The writing of this book would have remained a daydream were it not for the support of my agent, Jessica Papin, at the Dystel, Goderich & Bourret literary agency. I owe a lot to Jessica for her enthusiastic reception of the proposal right from the start, and without her this book would not be possible. (My zeal for gerund verbs has been significantly curtailed thanks to Jessica's input.) I would also like to thank my wonderful editors at St. Martin's Press, Sarah Grill and Charles Spicer, who had a vision for what this book could be, and helped shape it into what it is. Despite Sarah's contention that she learned a lot from me in reading and rereading the manuscript, in fact I learned a lot more from her, greatly elevating my writing skills. She helped keep me on track, and away from the scientific meanderings to which I was prone.

I also wish to thank the friends and colleagues, including Drs. Robert Bridges, Hector Rasgado-Flores, Pat McCormack, and Bonnie Blazer-Yost, who took time to review the science in the book. I thank them for their diligence, though any errors or omissions in the science remain entirely mine. Thanks also to my biochemistry professors at the University of St. Andrews in Scotland for initiating my interest in poisons, though I am sure that was not the intent of the course. I am sure many of the experiments we performed as undergraduates, including those with cyanide, are probably not allowed any longer. Finally, I would like to thank all the students who have passed through my classes during my years of teaching. You have always given an enthusiastic reception to my use of murder as a means to understand physiology. I am grateful for the privilege of helping you in your own journey of discovery. If I have omitted anyone from my thanks, I apologize but remain grateful—if embarrassed.

Notes

1. Gyles Brandreth, "How to commit the perfect murder," interview of Sir John Mortimer, *Telegraph* (London), December 18, 2001.

INTRODUCTION

1. The most notorious supplier of *aqua tofana* was a woman called Tofana of Sicily. She sold the potion as "Manna of St. Nicholas," after some holy water available at the time. Though it was marketed as a cosmetic, many purchasers seemed to use it as a poison, referring to it as *aqua tofana*. Some five hundred people have been estimated to have died because of its use. The governor of Naples finally ended the trade when he discovered that the poison, marketed in his city as Aquetta di Napoli, was so dangerous that six drops in a glass of wine would kill

the drinker. Tofana was arrested and after her confession was strangled to death in 1709. It is perhaps fortunate that the recipe for *aqua tofana* has not survived.

1. INSULIN AND MRS. BARLOW'S BATHTUB

1. There are two major types of diabetes. Type 1 is defined by the body's inability to make insulin, and type 2 is defined by the body's resistance to insulin or to insufficient production of insulin. Type 1 diabetes is also referred to as insulin-dependent diabetes, or juvenile diabetes, since this form mostly affects children, teenagers, and young adults.
2. All of Allen's scientific publications were little more than collections of anecdotes. A friend of Allen's noted that his manuscripts, all written in longhand, were mostly illegible to publishers, and he had to get his father to pay Harvard University for their publication. There is no doubt that starvation reduces glucose levels in patients with diabetes, but prolonged calorie restriction has its own problems, most obviously death from starvation, which Allen and Joslin euphemistically referred to as "inanition." But Joslin was not without compassion, commenting: "We literally starved the child and adult with the faint hope that something new in treatment would appear. . . . It was no fun to starve a child to let him live."
3. Insulin was discovered by Frederick Banting and Charles Best, but its first production also involved James Collip and John Macleod. Unfortunately the amazing story of insulin's discovery was marred by scientific jealousy, intense business compe-

tition, and even fistfights in the laboratory. Despite all four scientists being involved in testing insulin, only Banting and Macleod were given the Nobel Prize for insulin's discovery. Banting and Collip sold the patent for insulin to the University of Toronto for one Canadian dollar.

4. Although humans cannot digest fiber, it is still important in helping maintain normal intestinal function, and useful in preventing intestinal problems. Cows, like humans, do not possess enzymes to break down fiber, but specialized bacteria that grow in cows' intestines are able to digest fiber.
5. Nobel Prize–winner and mathematician John Nash, who won the Nobel Prize for Economics in 1994, suffered from schizophrenia, and was subjected to insulin shock therapy to treat his illness. His life story, and his insulin therapy, were portrayed in the 2001 movie *A Beautiful Mind.*
6. "My supposition was that some noxious agent weakened the resilience and metabolism of the nerve cells . . . a reduction in the energy of the cell, that is in invoking a minor or greater hibernation in it, by blocking off the cell with insulin will force it to conserve functional energy and store it to be available for the reinforcement of the cell." (M. Sakel, "The methodical use of hypoglycemia in the treatment of psychoses," reproduced in: *Am J Psychiatry* 151, supp. 6 [June 1994]: 240–247.)
7. The British medical journal *The Lancet* published the results of a controlled randomized clinical trial in which patients were either rendered unconscious by insulin treatment, or were given barbiturates to induce unconsciousness. There was no difference in outcome between the two groups, leading scientists to

conclude that whatever clinical benefits may ensue from a coma regimen, insulin was not the therapeutic agent.

8. The Home Office is the ministerial department of the British government responsible for, among other things, law and order and policing.
9. A fifty-nine-year-old cardiac surgeon was murdered when his wife replaced the insulin in his pump with etomidate, an anesthetic, and laudanosine, a muscle relaxant with similar effects to curare, causing the patient to stop breathing. The victim's wife was a nurse who worked in the recovery room of a nearby hospital and had easy access to the drugs. (B. Benedict, R. Keyes, and F. C. Sauls, *American Journal of Forensic Medicine and Pathology* 25 [2004]: 159–160.)

2. ATROPINE AND ALEXANDRA'S TONIC

1. During the French Revolution, red caps were worn by the citizens of Paris as a sign of support for the Revolution. Watching the heads of French aristocrats being lopped off by the guillotine was undoubtedly an amusing pastime, but people still had to eat—one ardent chef going so far as to suggest that the revolutionaries eat only red food. At the time tomatoes were not popular among the aristocracy, making them the perfect emblem for the bloodthirsty masses.
2. One of the earliest references to atropine in the English language comes from the medieval English botanist John Gerard. Gerard warned of the danger of deadly nightshade, which he called "sleepy" nightshade, and which he knew could be fatal.

He wrote "This kind of Nightshade causeth sleepe. . . . It bringeth such as have eaten thereof into a ded spleepe wherein many have died."

3. Although Geiger and Hess published the first description of how to purify atropine, there is some evidence that a German apothecary called Heinrich Mein may have produced atropine two years earlier in 1831.
4. One of the first uses of Marconi's wireless radio was to track the escape of murderer Dr. Harvey Hawley Crippen as he fled across the Atlantic Ocean. Unknown to Crippen, the entire world was following his escape. When he made land, Crippen was arrested by a Scotland Yard detective who had followed him on another ocean liner. Crippen had murdered his wife, Cora, by poisoning her with a chemical relative of atropine, the drug hyoscine. Crippen was hanged in Pentonville Prison in London in 1910.
5. The story of Loewi's somnambulant experiments may well be somewhat apocryphal. Loewi claimed to have had his dream about frog hearts on Easter weekend in 1920, however the journal in which he published his research received the manuscript the week before Easter of that year. Loewi, who liked telling stories, perhaps embellished the tale to increase the drama of the narrative. Nevertheless, the tale has survived down the years, no doubt to be retold in bedtime stories to the children of neuroscientists.
6. Among the deported Russian agents was Manhattan socialite, media personality, model, diplomat's daughter, and spy, Anna Chapman.

3. STRYCHNINE AND THE LAMBETH POISONER

1. In 1896 medical student Leonard Sandall, who had taken some strychnine as a tonic to help him through his exams, wrote in a letter to the medical journal *The Lancet*:

   > Three years ago, I was [studying] for an examination, and feeling "run down," I took 10 minims of strychnia solution. . . . On the second day of taking it, toward the evening, I felt a tightness in the "facial muscles" and a peculiar metallic taste in the mouth. There was great uneasiness and restlessness, and I felt a desire to walk about and do something rather than sit still and read. I lay on the bed and the calf muscles began to stiffen and "jerk." My toes drew up under my feet, and as I moved or turned my head flashes of light kept darting across my eyes. I then knew something serious was developing. ("An Overdose of Strychnine," *The Lancet* 147 [1896]: 887.)

2. "*The Mysterious Affair at Styles*," *Pharmaceutical Journal and Pharmacist* 57 (1923): 61. In fact, Christie was a trained and certified pharmacist during World War I, having passed her apothecary examinations in 1917.
3. Although strychnine held the record for being the bitterest substance known, that notoriety was to fade in 1955, when Friedhelm Korte, at the University of Hamburg, isolated a plant chemical from the gentian species that he called amarogentin. The bitter taste of amarogentin can be detected at a remarkable

dilution of one part in 58 million (meaning if you put a drop of amarogentin in an Olympic-size swimming pool you could still taste it), making it roughly one thousand times more bitter than strychnine.

4. Public hangings had been outlawed since 1868, and prisoners were now executed within the confines of the prison walls. However, the crowds gathering outside Newgate Prison as Cream was led to the scaffold were so large—and all clamoring for his death—that one newspaper wrote: "Probably no criminal was ever executed in London who had a less pitying mob awaiting his execution." At his hanging, Cream's last words, as reported by his executioner James Billington, were "I am Jack the—." Although Billington maintained he had hanged Jack the Ripper, Cream's incarceration in Illinois at the time of the Ripper murders argues against it. Such was Cream's notoriety though, that immediately after his hanging Madame Tussaud's waxworks paid £200 for his clothes and personal effects to place on his effigy.
5. From the transcript of testimony given by Matilda Clover's landlady during Cream's trial at the Old Bailey.
6. Agatha Christie, *The Mysterious Affair at Styles*. London: The Bodley Head, 1921.

4. ACONITE AND MRS. SINGH'S CURRY

1. Recollections of Lamson's discussions on Percy's pill taking were noted by His Majesty's Coroner for the City of London and the Burrough of Southwark, during Bedbrook's testimony

at Lamson's trial. The comments were subsequently reported by the coroner, F. J. Waldo, in "Notes on Some Remarkable British Cases of Criminal Poisoning," *Medical Brief* 32 (1904): 936–940.

2. "The Case of Poisoning at Wimbledon," *Pharmaceutical Journal and Transactions* 12 (1881–1882): 777–780.
3. Trial transcript in *Trial of John Hendrickson Jr. for the Murder of His Wife Maria by Poisoning*. Albany, NY: Weed Parsons and Co. Printers, 1853.
4. So-called witches' salve was made from aconite dissolved in fat and rubbed on the body so that the aconite was slowly absorbed through the skin. The sensation of limbs floating, and of being cut off from the ground, was likely the basis for witches' presumed ability to fly.
5. The attending physicians wrote up their notes as a case report for publication. (K. Bonnici, et al., "Flowers of Evil," *The Lancet* 376, no. 9752 [2010]: 1616.)

5. RICIN AND GEORGI'S WATERLOO SUNSET

1. In fact, a later careful examination of the X-ray film did show a tiny pellet embedded in Markov's leg, but the radiologist assumed it was a defect in the film since it was so small.
2. Vladimir Kostov had also defected from Bulgaria and was working for Radio Free Europe in Paris. On August 27, 1978, Kostov was attacked on an escalator at the Arc de Triomphe Métro station. As he neared the top of the escalator, Kostov felt a prick in the small of his back. He turned to see a man

carrying a briefcase. The next day Kostov developed a fever and a swelling around the injection site. After Markov's death, Kostov allowed doctors to remove tissue from around where a pellet had lodged; the pellet contained ricin. Kostov survived because the wax coating on the pellet failed to dissolve fully, and most of the ricin remained within the pellet. The slow leak of ricin was sufficient for Kostov's body to mount an immune response, developing antibodies that neutralized the toxin.

3. Georgi Markov, *The Truth That Killed.* New York: Ticknor and Fields, 1984.

6. DIGOXIN AND THE ANGEL OF DEATH

1. Foxglove first appears in Old English as the same, "foxglove," so it cannot be a name that has been altered or corrupted over time. Indeed, one of the first historical references on digitalis is a manuscript copy of the *Herbarium Apulueii Platonica*, written in Bury St. Edmunds, England, around 1120, where the word *foxglove* appears as it has for the last one thousand years.
2. Telephone conversation between Dr. Steven Marcus, poison control center director, and Dr. William Cors, medical director:

   Marcus: This is a police matter.

   Cors: What we're wrestling with is, you know, throwing the whole institution into chaos, versus, you know, responsibility to, you know, protect patients from further harm.

> And we have been trying to investigate this to get some more information before we made any kind of rush to, you know, judgment.

"Angel of Death" *60 Minutes,* CBS, April 28, 2013.

7. CYANIDE AND THE PROFESSOR FROM PITTSBURGH

1. Dippel, who was born and grew up in Castle Frankenstein (later to become the site of Mary Shelley's most famous novel), studied Theology and Alchemy at the University of Göttingen. During his studies one of his professors described him as one "whose brain seems to have been heated to a high degree of fermentation by the heat of the laboratory."
2. In 1860 Professor Robert Christison (see chapter 4) received a letter from the captain of a whaling ship based at the Scottish city of Leith. The captain inquired whether using a capsule of hydrocyanic acid attached to the harpoon would be effective in bringing about the death of a whale. Christison replied that it would certainly work, and a few "successful" trials were undertaken. But once the crews saw the effects of cyanide on something as large as a whale, they were understandably reluctant to cut up or even touch the massive body. At the time whaling was a major industry, with ships setting sail from many ports on the East Coast of Scotland to provide whale oil for lamps. Christison, who also dabbled as a research chemist, did some pioneering work on the nature of paraffins, work that was to lead to making whale oil obsolete.

3. The idea that humans can directly access energy from the sun has been promulgated in the pseudo-science of breatharianism. A 2010 Australian documentary told the story of an Indian guru who claimed to have lived off sunlight and without food or water for seventy years. Unfortunately such nonsense was taken seriously by a fifty-year-old Swiss woman who took on the diet of sunlight and air. Predictably this did not end well, and the woman was found dead in her home in the Swiss town of Wolfhalden.
4. In the 1890s one audacious, or foolhardy, physician swallowed a small dose of potassium cyanide in order to determine the effects of cyanide on the human body. The medical journals described his gasping cries that he "was suffocating." Although the doctor did survive his self-inflicted ordeal, it is a testament to common sense that no one has repeated his experiment.
5. Taken from court transcript of Ferrante's 911 call:

911: Allegheny County 911. What's the address of your emergency?

Ferrante: Hello. Please, please, please. I'm at 219 Lytton Avenue. I think my wife is having a stroke.

911: Is she able to talk at all?

Ferrante: No, she's not saying a word. Now, now she's like having a seizure like she's [groaning]. Her eyes are still open. She's looking, she just closed them. [groaning] Oh God, help me. God help me.

911: OK, like I said, I'm sending the paramedics to help you. OK. I'm making sure help is on the way. Don't let her have anything to eat or drink OK, Bob?

Ferrante: Oh God, help me.

911: OK Bob, OK don't let her have anything to eat or drink or make her suck and cause problems for the doctor OK. Just, I want you to let her rest in a comfortable position and wait for help to arrive.

Ferrante: Her, her, her folks are down at Shadyside, maybe that would be the best place to take her.

911: OK let the paramedics know that when they arrive that you want to have her taken to Shadyside, OK?

Ferrante: I, I, I will.

8. POTASSIUM AND THE NIGHTMARE NURSE

1. Salt substitutes are 60 percent potassium chloride and 40 percent sodium chloride.
2. According to Dan Koeppel in his 2008 book *Banana*, Americans eat more bananas per year than apples and oranges com-

bined. Bananas are not actually a fruit, but rather a berry, and the banana tree is not a tree, it's a large herb. (Dan Koeppel, *Banana: The Fate of the Fruit that Changed the World.* New York: Plume, 2008, xi.

3. In her 1951 book *Let's Have Healthy Children,* nutritionist Adelle Davis argued that colicky babies could be soothed with potassium chloride. One mother of a two-month-old took this advice to heart and mixed 3,000 mg of potassium chloride with her breast milk and fed her baby. The next morning, in the same manner, 1,500 mg of potassium chloride was given. A few hours later the baby was listless, had turned blue, and finally stopped breathing. The child was rushed to hospital but died two days later. The child's blood potassium level was three times higher than normal. Despite heavy criticism by peers for her recommendations, which were not based on any scientific evidence, Davis remained hugely popular among parents in the 1960s.

9. POLONIUM AND SASHA'S INDISCRIMINATE INTESTINE

1. In 1970 the USSR sent the lunar rover Lunokhod I to the moon's surface (no one has retrieved it yet), where its electronic components were kept warm by the radioactive decay of polonium-210.
2. In 1960 U.S. pilot Francis Gary Powers was photographing the Mayak facility when his CIA spy plane was shot down.
3. Press conference held in Moscow, November 17, 1998.
4. At the time of Litvinenko's assassination, the U.S. Embassy was in the London Chancery Building at 24 Grosvenor Square. The

embassy opened at its new location, south of the Thames River, in Nine Elms, Battersea, in January 2018. Designed by architect Kieran Timberlake, the building is meant to look like a crystalline cube. Reagan's statue remains in Grosvenor Square.

5. https://assets.publishing.service.gov.uk/government/uploads/system/uploads/attachment_data/file/493860/The-Litvinenko-Inquiry-H-C-695-web.pdf

10. ARSENIC AND MONSIEUR L'ANGELIER'S COCOA

1. The court of Louis XIV, the Sun King, was rocked by the "Affair of the Poisons." Salacious tales of love potions, witchcraft, and murder enthralled all of Europe. At the heart of the saga was Catherine Deshayes Monvoisin, known as La Voisin. When her husband went bankrupt, La Voisin made her fortune performing abortions and providing love potions and deadly poisons in equal measure. La Voisin was also consulted by Madame de Montespan, the king's mistress, who sought to regain the king's favor using Voisin's aphrodisiacs. La Voisin was eventually arrested and executed for witchcraft.
2. The act, passed in 1851, severely restricted who could buy arsenic, but there were no restrictions on who could sell it. Until 1868 there was no legal definition of a pharmacist, and when legal pharmacies were created, all pharmacists had to keep a written register of who purchased arsenic. The maximum penalty for breaching the provisions of the act, or providing false information, was £20, roughly £3,000 or $4,000 today.

3. The Styrian defense was used with varying degrees of success at the trials of Madeleine Smith in 1857 and Florence Maybrick in 1889.

11. CHLORINE AND THE KILLER NURSE OF LUFKIN

1. Haber had hoped to shorten the war by killing and demoralizing Allied troops in the trenches. His hopes were wildly off the mark, as the war ground on for another three and a half years.
2. Ignaz Semmelweis, *The Etiology, Concept, and Prophylaxis of Childbed Fever* (1861). Translated and edited by K. Codell Carter. Madison: University of Wisconsin Press, 1983.
3. Bleach, or more specifically the chlorine in bleach, is easily detected by a reaction that causes a color change. DPD (N,N-diethyl-p-phenylenediamine) is colorless until it reacts with chlorine and changes color. The more intense the color, the more chlorine is present. DPN can be used as drops or impregnated into paper to make test strips.
4. When white blood cells encounter bacteria, they can kill them using bleach made by the cells themselves. When this happens, the chlorine in bleach can attach itself to tyrosine, one of the twenty amino acids that make up protein. The presence of chlorotyrosine can be used to monitor the body's response to an infection, since small amounts of chlorotyrosine can be measured. In the case of Kimberly Saenz, the amount of chlorotyrosine detected was several hundred times greater than would ever be found during an infection.

Selected Bibliography

Below is a selection of interesting books and articles that provided some of the source material for topics in this book.

GENERAL

Blum, D. *The Poisoner's Handbook: Murder and the Birth of Forensic Medicine in Jazz Age New York.* New York: Penguin Books, 2010.

Christison, R. A. *A Treatise on Poisons in Relation to Medical Jurisprudence, Physiology and the Practice of Physic.* Edinburgh: John Stark, 1829.

Emsley, J. *The Elements of Murder.* Oxford: Oxford University Press, 2005.

Evans, C. *The Casebook of Forensic Detection.* New York: John Wiley & Sons, 1996.

Farrell, M. *Poisons and Poisoners: An Encyclopedia of Homicidal Poisons.* London: Bantam Books, 1994.

Gerald, M. C. *The Poisonous Pen of Agatha Christie.* Austin: University of Texas Press, 1993.

Glaister, J. *The Power of Poison.* London: Christopher Johnson, 1954.

Harkup, K. *A Is for Arsenic: The Poisons of Agatha Christie.* New York: Bloomsbury, 2015.

Herman, E. *The Royal Art of Poison: Filthy Palaces, Fatal Cosmetics, Deadly Medicine, and Murder Most Foul.* New York: St. Martin's Press, 2018.

Holstege, C. P., et al. *Criminal Poisoning: Clinical and Forensic Perpectives.* Burlington, MA: Jones & Bartlett Learning, 2010.

Johll, M. E. *Investigating Chemistry: A Forensic Science Perspective.* New York: Freeman and Co., 2007.

Macinnis, P. *Poisons from Hemlock to Botox and the Killer Bean of Calabar.* New York: Arcade Publishing, 2004.

Mann, J. *Murder, Magic and Medicine.* Oxford: Oxford University Press, 2000.

McLaughlin, T. *The Coward's Weapon.* London: Robert Hales, 1980.

Ottoboni, M. A. *The Dose Makes the Poison.* New York: Van Nostrand Reinhold, 1991.

Reader, J. *Potato: A History of the Propitious Esculent.* New Haven: Yale University Press, 2009.

Stevens, S. D., and A. Klarner. *Deadly Doses: A Writer's Guide to Poisons.* Cincinnati: Writer's Digest Books, 1990.

Thompson, C. J. S. *Poisons and Poisoners.* London: Harold Shaylor, 1931.

Trestrail, J. H. III. *Criminal Poisoning.* Totowa, NJ: Humana Press, 2007.

1. INSULIN

Ackner, B., A. Harris, and A. J. Oldham. "Insulin Treatment of Schizophrenia: A Controlled Study," *The Lancet* 272, no. 6969 (1957): 607–611.

Allen, F. "Studies Concerning Diabetes," *JAMA* 63 (1914): 939–943.

Askill, J., and M. Sharpe. *Angel of Death.* London: Michael O'Mara Books,1993.

Bathhurst, M. E., and D. E. Price. "Regina v Kenneth Barlow," *Med. Leg. J.* 26 (1958): 58–71.

Bliss, M. *The Discovery of Insulin.* Chicago: University of Chicago Press, 2007.

Bourne, H. *The Insulin Myth. The Lancet* 263 (1953) : 48–49.

Joslin, E. "The Diabetic," *Journal of the Canadian Medical Association* 48 (1943): 488–497.

Marks, V., and C. Richmond. *Insulin Murders—True Life Cases.* London: Royal Society of Medicine Press, 2007.

Parris, J. *Killer Nurse Beverly Allitt.* Scotts Valley, CA: CreateSpace Independent Publishing, 2017.

Peterhoff, M., et al. "Inhibition of Insulin Secretion via Distinct Signaling Pathways in Alpha2-Adrenoceptor Knockout Mice," *Eur. J. Endocrinol.* 149 (2003): 343–350.

2. ATROPINE

Carter, A. J. "Narcosis and Nightshade," *British Medical Journal* 313 (1996):1630–1632.

Christie, A. "The Thumb Mark of St. Peter," In *The Thirteen Problems*. Glasgow: Collins Crime Club, 1932.

Harley, J. *The Old Vegetable Neurotics: Hemlock, Opium, Belladonna and Henbane.* Charleston, NC: Nabu Press, 2012.

Holzman, R. S. "The Legacy of Atropos, the Fate Who Cut the Thread of Life," *Anesthesiology* 89 (1998): 241.

Marcum, J. A. "'Soups' vs. 'Sparks': Alexander Forbes and the Synaptic Transmission Controversy," *Annals of Science* 63 (2006): 638.

People vs. Buchanan, Court of Appeals of the State of New York, 145 N.Y.1 (1895).

3. STRYCHNINE

Bates, S. *The Poisoner: The Life and Crimes of Victorian England's Most Notorious Doctor*. London: Duckworth Press, 2014.

Buckingham, J. *Bitter Nemesis: The Intimate History of Strychnine*. Boca Raton, FL: CRC Press, 2008.

Graves, R. *They Hanged My Saintly Billy: The Life and Death of Dr. William Palmer*. Garden City, NY: Doubleday, 1957.

Griffiths-Jones, A. J. *Prisoner 4374*. London: Macauley Publishers Ltd., 2017.

Li, W-C, and P. R. Moult. "The Control of Locomotor Frequency by Excitation and Inhibition," *J. Neurosci.* 32 (2012): 6220–6230.

Matthews, G. R. *America's First Olympics: The St. Louis Games of 1904*. Columbia: University of Missouri Press, 2005.

4. ACONITE

American Medicine 5, "Of Poisons and Poisonings," editorial comment (June 20, 1903): 977.

Headland, F. W. "On Poisoning by the Root of Aconitum nepellus," *The Lancet* 1 (1856): 340–343.

Turnbull. A. *On the Medical Properties of the Natural Order Ranunculaceae: And More Particularly on the Uses of Sabadilla Seeds and Delphinium Straphisagria*. Philadelphia: Haswell, Barrington and Haswell 1838.

Wells, D. A. "Poisoning by Aconite: A Second Review of the Trial of John Hendrickson Jr.," *Medical and Surgical Reporter (Philadelphia)* (1862): 110–118.

5. RICIN

Ball, P. *Murder under the Microscope*. London: MacDonald, 1990.

Markov, G. *The Truth That Killed*. London: Littlehampton Books, 1983.

Schwarcz, J. *Let Them Eat Flax*. Toronto: ECW Press, 2005.

6. DIGOXIN

Graeber, C. *The Good Nurse: A True Story of Medicine, Madness, and Murder.* New York: Hachette Book Group, 2013.

Kwon, K. "Digitalis Toxicity," eMedicine, July 14, 2006; www.emedicine.com/ped/topic590.htm

Olsen, J. *Hastened to the Grave.* New York: St. Martin's Paperbacks, 1998.

Withering, W. *An Account of the Foxglove and Some of Its Medicinal Uses.* London: G.G. and J. Robinson, 1785.

7. CYANIDE

Christison, R. "On the Capture of Whales by Means of Poison," *Proc. Roy. Soc. Edin.* iv (1860): 270–271.

Gettler, A. O., and A. V. St. George. "Cyanide Poisoning," *American Journal of Clinical Pathology* 4 (1934): 429.

Hunter, D. *Diseases of Occupations.* London: Hodder & Stoughton, 1976.

Kirk, R. L., and N. S. Stenouse. "Ability to Smell Solutions of Potassium Cyanide," *Nature* 171 (1953): 698–699.

Ward, P. R. *Death by Cyanide: The Murder of Dr. Autumn Klein.* Lebanon, NH: University Press of New England, 2016.

8. POTASSIUM

Anderson, A. J., and A. L. Harvey. "Effects of the Potassium Channel Blocking Dendrotoxins on Acetylcholine Release and Motor Nerve Terminal Activity," *Br. J. Pharmacol.* 93 (1988): 215.

Ebadi, S., with A. Moaveni. *Iran Awakening.* New York: Random House, 2006.

Koeppel, Dan. *Banana: The Fate of the Fruit that Changed the World.* New York: Plume, 2008.

Manners, T. *Deadlier Than the Male.* London: Pan Books, 1995.

Webb, E. *Angels of Death: Doctors and Nurses Who Kill.* Victoria, Australia: The Five Mile Press, 2019.

9. POLONIUM

Brennan, M., and R. Cantrill "Aminolevulinic Acid Is a Potent Agonist for GABA Autoreceptors," *Nature* 280 (1979): 514–515.

Emsley, J. *Elements of Murder,* Oxford: Oxford University Press, 2005.

———. *Molecules of Murder.* Cambridge: Royal Society of Chemistry, 2008.

Harding, L. *A Very Expensive Poison.* New York: Vintage Books, 2016.

Owen, R. "The Litvinenko Inquiry" (2016). https://assets.publishing.service.gov.uk/government/uploads/system/uploads/attachment_data/file/493860/The-Litvinenko-Inquiry-H-C-695-web.pdf

Quinn, S. *Marie Curie: A Life.* Cambridge, MA: Perseus Books, 1995.

Sixsmith, M. *The Litvinenko File.* London: Macmillan, 2007.

10. ARSENIC

Blum, D. *The Poisoner's Handbook.* New York: Penguin Books, 2010.

Cooper, G. *Poison Widows: A True Story of Witchcraft, Arsenic and Murder.* London: St. Martin's Press, 1999.

Fyfe, G. M., and B. W. Anderson. "Outbreak of Acute Arsenical Poisoning," *The Lancet* 242 (1943), 614–615.

Goyer, R. A., and T. W. Clarkson. *Toxic Effects of Metals: The Basic Science of Poisons.* New York: McGraw-Hill, 2001.

Livingston, J. D. *Arsenic and Clam Chowder: Murder in Gilded Age New York.* Albany: SUNY Press, 2010.

Parascandola, J. *King of Poisons: A History of Arsenic.* Lincoln, NE: Potomac Books, 2012.

Vahidnia, A., G. B. van der Voet, and F. A. de Wolf. "Arsenic Neurotoxicity—A review," *Human and Experimental Toxicology* 26 (2007): 823.

Whorton, J. C. *The Arsenic Century: How Victorian Britain Was Poisoned at Home, Work, and Play*. New York: Oxford University Press, 2010.

11. CHLORINE

Foxjohn, J. *Killer Nurse*. New York: Berkley Books, 2013.

Hurst, A. *Medical Diseases of the War* (1916). Plano, TX: Wilding Press. 2009.

Keegan, J. *The First World War*. New York: Vintage Books, 1999.

Saenz v. State of Texas, court report, www.courtlistener.com/opinion/4269367/kimberly-clark-saenz-v-state/

Pressage Professional Inc.

Neil Bradbury, Ph.D., is a professor of physiology and biophysics at the Rosalind Franklin University of Medicine and Science, where he teaches and conducts research on genetic diseases. A full-time scientist and educator, Bradbury has won numerous awards for his unique approach to teaching physiology. He has presented his research around the world and authored more than eighty scientific articles and book chapters. He currently lives in Illinois with his wife and two border collies. *A Taste for Poison* is his first book.